ACCLAIM FOR

Out of Eden

"Combining personal meditation, travel narrative, and excellent reportage, *Out of Eden* creates a rich and panoramic view of life on our planet." —Alan Lightman, author of *Einstein's Dreams*

"Alan Burdick is my new favorite writer. I love the quiet poetry of his prose; his pitch-perfect wit; and his calm, potent mastery of the facts. I actually stop and think, 'Wow!' as I'm reading."
—Mary Roach, author of *Spook*

"Graceful and inviting."
—Richard Conniff, *The New York Times Book Review*

"A wonderful book for anyone interested in the least about the mysteries of ecological dynamism, our considerable role in shaping it and often lame attempts to control it . . . [Alan Burdick] writes with graceful simplicity." —Richard Seven, *The Seattle Times*

"[An] intelligent and thought-provoking book . . . A captivating odyssey, a physical and intellectual journey."
—Alcestic "Cooky" Oberg, *Houston Chronicle*

"Burdick has a gift for compacting large quantities of time into a handful of trim, lovely sentences . . . This is what good books are about."
—Anthony Doerr, *The Boston Globe*

"His tour through the burgeoning discipline of invasion ecology is nuanced, judicious and often delightful; in the finest tradition of science writing."
—Andrew O'Hehir, *Salon*

"Highly readable and thought-provoking." —Allan Watt, *New Scientist*

"Burdick's fascination with the science is contagious . . . This is a captivating book with wide-ranging appeal."
—*Publishers Weekly* (starred review)

"A silkily written excursion into the evolution of ecosystems and the possible threats to biodiversity from newcomers." —*Kirkus Reviews*

Alan Burdick

Out of Eden

ALAN BURDICK's articles and essays have appeared in *The New York Times Magazine*, *Best American Science and Nature Writing*, *Harper's*, *GQ*, *Natural History*, and *Discover*, where he is a senior editor. He lives with his wife in New York City.

Out of Eden

Out of Eden

An Odyssey of Ecological Invasion

Alan Burdick

Farrar, Straus and Giroux / New York

Farrar, Straus and Giroux
19 Union Square West, New York 10003

Distributed in Canada by Douglas & McIntyre Ltd.
Printed in the United States of America
Published in 2005 by Farrar, Straus and Giroux
First paperback edition, 2006

Portions of this book have appeared, in different form, in Discover *and* The New York Times Magazine.

Grateful acknowledgment is made to the following for permission to reprint the images in this book: Dean Amadon, "The Hawaiian Honeycreepers (Aves, Drepaniidae)," Bulletin of the American Museum of Natural History: *New York, 1950, volume 95, Article 4; C. B. Huffaker, "Experimental Studies of Predation: Dispersion Factors and Predator-Prey Oscillations,"* Hilgardia 27: *343–83, copyright © 1958 Regents of University of California; "Copepod," National Marine Invasions Laboratory, Smithsonian Environmental Research Center.*

Library of Congress Control Number: 2005922517

Paperback ISBN-13: 978-0-374-53043-3
Paperback ISBN-10: 0-374-53043-2

Designed by Jonathan D. Lippincott

www.fsgbooks.com

3 4 5 6 7 8 9 10

To Mary and Robert,
for the roots

Why, Cadmus, why stare at the snake you've slain?
You too shall be a snake and stared at.

—Ovid, *Metamorphosis*

Out of Eden

Flight

Out of the blue, a red fuse: Hawaii.

The first island erupted from a volcanic vent in the seabed, drifted northwest with the ocean crust, and sank. Called the Meiji Seamount, it is more than eighty million years old and today lies at the bottom of the Pacific Ocean not far from Russia's Kamchatka Peninsula and the Aleutian Islands, buried under half a mile of sea clay, chalk, and a restless film of biological detritus known to scientists as "ooze." Eons passed; more islands came, went, stayed. An archipelago formed of the semisubmerged—a line of stepping-stones pointing back to the beginning. Upon them, incidental travelers settled: a spider ballooning across from the mainland on a strand of silk, a snail or a tick or a burr stowed in the down of a roving seabird, each new colonist arriving every twenty thousand years or so. Natural selection did its pruning. A pair of finches gave rise, over millions of years, to fifty-odd radiant species of honeycreeper. From a single pair of fruit flies, more than six hundred species evolved. Thornless raspberries, nettleless nettles, cave-dwelling albino crickets. No reptiles, no mosquitoes, no mammals, except, eventually, a monk seal and an insectivorous bat. The meek shall inherit paradise.

Out of the blue—restless hours of it, even in this age of air travel—Hawaii. I watched the archipelago unfurl below my window, east to west. First the Big Island, Hawai'i with apostrophe. Clouds wreathed its peaks, Mauna Loa and Mauna Kea, at ten thousand feet; one million years old, it is the youngest member of the island chain. Twenty minutes later, Maui and the sprawling crater of Haleakala, which residents have come to believe is an extinct volcano but in fact is merely dormant. Then Molokai and Lanai, low and rippled as green flatworms. In the passenger cabin, on the movie screen at the front of the aisle, a map of the island chain appeared suddenly. We were the small white silhouette of a jet arc-ing toward Honolulu, which was marked as a small red star on the island of Oahu.

A stewardess came down the aisle handing out slips of paper printed with orange lettering: declaration forms. I declare: No, I harbor no fresh fruits or vegetables. I have no live lobsters or clams to speak of, no flowers, foliage, rooted plants, or plant cuttings; no seeds, bulbs, soil, or sand; no bacteria, no algae, no fungi, no protozoa. No, I am not traveling with a dog or a bird or a turtle or a lizard. Oahu appeared through the porthole: Diamond Head crater to the west, the high-rises of Waikiki merging into those of Honolulu, pearls of concrete strung along a white sand beach. The plane banked over sugarcane fields, over Pearl Harbor. With the continents of Asia and North America more than two thousand miles away, Hawaii is the most isolated major landmass on Earth, the pilot announced. His voice rang over the loudspeaker: *We are farther from anywhere than anywhere.*

I had come, ostensibly, to see about a snake. That is an uncommon pastime in Hawaii, for the simple reason that, as far as anyone can yet determine, there are few snakes to be found. Once or twice a year someone's pet boa constrictor or Burmese python escapes and reappears in a Waikiki or Waimea garage, prompting a call to the animal squad at the Hawaii Department of Agriculture. The only snake known to be established in the state is the Braminy blind snake, a sightless, wriggling creature closer in size and spirit to a worm.

The snake I sought was, like me, a stranger to Hawaii: *Boiga irregularis*, the brown tree snake. Originally from Australia and Indonesia, the snake arrived first on the Pacific island of Guam, thirty-three hundred miles west of Honolulu, shortly after the Second World War. The snake's sphere of influence and notoriety has expanded steadily since. Prior to that fateful arrival, the only snake on Guam was the hapless Braminy. Today Guam hosts more brown tree snakes—more snakes of any kind, for that matter—per square mile than anywhere else in the world. This distinction has come largely at the expense of Guam's native bird population, which the snake's boundless appetite has almost entirely eliminated; the national bird of Guam, a flightless rail known as the koko, reigns from within the safety of a snake-proof pen in a rearing compound near the international airport.

In its native territory, the brown tree snake rarely grows more than three feet long; on Guam, twelve-foot-long specimens are not unknown. Its bite, inflicted on two hundred or so people a year, is slightly venomous, akin to a bee sting. It also displays an irrepressible ability to climb, most notably onto power lines and into transformers, causing dozens of electrical outages each year, at a cost approaching a million dollars annually. Its toll on the local psyche is less easily quantified. Tales circulate of the snake that crawled in through the toilet; the snake that leaped from an automobile air-conditioning vent, sending the driver into a near-fatal swerve; and the snakes that attack infants in their cribs—lured there, local housewives attest, by the scent of mother's milk.

And now, it seems, the brown tree snake has gained the ability to fly. Since 1981, eight brown tree snakes have been found near the runways of Honolulu's airports. To the best that experts can determine, the snakes arrived as stowaways in the wheel wells of jetliners from Guam. A ninth specimen, last seen on the perimeter of the local air force base, also may have been a brown tree snake, but it slipped away before it could be positively identified. Over the centuries, Hawaii's bird population has absorbed assaults by one introduced organism after another: humans with clubs; avian malaria, which arrived with imported cage birds and was transmitted by an introduced mosquito; Norway rats, which arrived incidentally aboard European ships; and mongooses, which were introduced intentionally in the late nineteenth century with the aim—misguided, as it turned out—of controlling the rat population. Today, nearly 40 percent of the birds on the U.S. endangered species list are found in Hawaii, an indication both of the uniqueness of Hawaii's avian fauna and of its precarious situation. None of those birds are evolutionarily prepared for the likes of snakes.

Nor, presumably, are the eighteen thousand human visitors who arrive in Hawaii daily. A representative fraction of us had begun to gather in baggage claim: honeymooners, Japanese businessmen, grandmothers in leis, flight-weary tourists in floral print T-shirts. We had received a warning of sorts on the incoming flight. For a few moments, the map of Hawaii on the screen in the passenger cabin had given way to a video called *It Came from Beyond*, made by the Hawaii Department of Agriculture. A local celebrity appeared. *Aloha!* Hawaii is a special place; thanks for visiting! But Hawaii has other, less welcome visitors: insects, animals, plants, and diseases that can threaten local agriculture or native wildlife if they become established. So be please be careful what you bring in. Better to declare it than to suffer a fine. Better to confess. In baggage claim I watched as an avid beagle, outfitted in a green vest marked HAWAII DEPARTMENT OF AGRICULTURE, towed an amiable state employee around the room by a leash, through a sea of suitcases, tote bags, backpacks, bundles, and their flagging possessors. The dog zigzagged onward, nosily intent, until at last he zeroed in on a small sensory paradise, a black handbag resting on the marble floor. The dog's inspector spoke briefly with the bag's owner, a Korean woman who eyed the beagle with terror. Reluctantly, she withdrew two oranges from the handbag and handed them to the inspector.

Eight brown tree snakes in twenty years would hardly seem to pose much of a threat. However, the biology of *Boiga irregularis* is sufficiently remarkable to concern any scientist, conservationist, or tourist administrator. The snake is nocturnal and, like any good predator, impressively hard to find. It is also impressively hardy; in 1993 a military officer in Corpus Christi, Texas, opened the lid of a washing machine he had packed up and sent home several months earlier from his previous station on Guam. Inside was a brown tree snake, alive, with nothing for company or nourishment but a small pool of water. Thomas Fritts, the coordinator of the Brown Tree Snake Research Program for the United States and Guam, kept one alive in his office for a full year without once feeding it. Like many reptiles, a female brown tree snake can lay fertile eggs for several years after mating. Two dozen eggs a year for seven years equals more than a hundred and fifty eggs, a hundred and fifty new snakes capable of hiding, mating, and producing yet more eggs—all from a single snake. On Guam, four decades passed before scientists realized that the island was thoroughly snake infested, its bird population doomed, its ecosystem permanently altered. Today in Hawaii there is similar, urgent wondering: Perhaps those eight snakes are mere harbingers. Perhaps they are but a visible handful of innumerable brown tree snakes that have slipped, and continue to slip, across Hawaii's borders and into its foliage—apparent warning signs of what in fact is an encroaching, multiplying multitude of snakes. Perhaps below the Edenic surface of bougainvillea-lined streets and moss-draped rain forests, there lies a swarming Hieronymus Bosch world of brown tree snakes slowly gathering into a wicked, irreversible mass.

"What havoc the introduction of any new beast of prey must cause in a country," Charles Darwin mused in 1835, "before the instincts of the indigenous inhabitants have become adapted to the stranger's craft or power." Darwin was twenty-six years old at the time, a passenger aboard the HMS *Beagle*, on what would prove to be a seminal journey through the eastern Pacific Ocean. Some years later he waxed enthusiastic for the possibilities in Hawaii—"I would subscribe 50 pounds to any collector to go there and work"—but his professional fate became entwined instead with the Galápagos Islands, six hundred miles off the coast of Ecuador. Darwin spent several weeks there collecting finches and comparing the shells of tortoises, noting the morphological differences between the species of one island and the next. This body of evidence would later fuel his theory of evolution by natural selection. A creature becomes isolated—a mainland finch or spider blown to a distant outpost, a tortoise marooned over eons by the sinking of a land bridge or the rise of an impassable mountain chain. The struggle to survive commences; competition ensues for limited food and mates. The winners survive to reproduce, and their descendants continue in the struggle. The losers die, leaving fewer offspring, a withering branch on the evolutionary tree. And so the limb is trimmed from generation to generation until a new branch, a new species, sufficiently distinct from the ancestral one, takes form.

That insight would come later. The immediate question on Darwin's mind was more straightforward: Why are certain animals and plants where they are and not somewhere else? Why are the inhabitants of this island similar to, yet recognizably different from, the inhabitants of that island? Why isn't the world's flora and fauna everywhere the same? What makes a place, and its residents, unique? "One of the subjects on which I have been experimenting & which cost me much trouble, is the means of distribution of all organic beings found on oceanic islands," he wrote to a colleague in 1857. The conclusions he presented, in the 1859 publication of *On the Origin of Species* (and the five subsequent revisions), unveiled the fabric of nature across geologic time and geographic space. Nature is dynamic, not static. Species that existed long ago exist no longer; new species have since arisen where no such species existed before. Things move around. Life, if left alone long enough, transforms itself. Eventually—so slowly that the human eye can detect it only well in retrospect—this place, in its flora and fauna, becomes different from that place. Heterogeneity arises and is continually reborn.

That variegated fabric, scientists today fear, is unraveling from within.

Now, as never before, exotic plants and organisms are traversing the globe, borne on the swelling tide of human traffic to places where nature never intended them to be. Africanized bees have reached California; stinging colonies of South American fire ants have settled in Texas; the kudzu vine is strangling the southeast; the zebra mussel, a pistachio-size mollusk from Europe, carpets the bottom of the Great Lakes and, increasingly, the Mississippi River, where it slurps the water clean of plankton that other aquatic creatures require to thrive. Among the invaders to grab headlines lately is an Asian fish called the snakehead. A delicacy to some, an ornamental fish to others, the snakehead gradually entered U.S. waterways as aquarium owners emptied their pets into local ponds and streams. The animal proliferated and its population spread, emerging recently in Maryland and Florida and alarming fish biologists with its voracious appetite for the local fauna. Efforts to eliminate it by poisoning entire ponds have failed; the animal merely burrows into the mud or, with amphibian-like versatility, crawls up onto land, where it can remain for days. Live specimens have been found on land miles from the nearest body of water; apparently, when one pond or river fails them, they walk to the next one.

The flight of the brown tree snake is merely one of the more dramatic steps toward what some experts in biological invasion have begun to refer to fearfully as "the homogenization of the world." Feral pigs now root in the lawns of San Jose, California. Giant Asian carp, introduced in the 1970s to control aquatic weeds, leap unsolicited into fishing boats along the Mississippi River. In New York City, where one is tempted to think that there's already one of everything, environmental officials are closely monitoring the advance of the Asian long-horned beetle, which so far has required the anguished removal of several city blocks' worth of maples from Brooklyn. At last report, two trees in Central Park were found to be infected; to check any further spread, researchers are experimenting with a stethoscope-like device that listens for the chewing of beetle maggots within the tree. All told, five thousand introduced plant species now exist in U.S. ecosystems, compared with the seventeen thousand known native plant species. Half the wild poisonous plants in North America are introduced, as are half the earthworms in the soil. In south Florida, the nexus of the nation's pet trade, backyard menageries are so common and escapes so commonplace that exotic pythons and boa constrictors are now established, free-ranging residents. The local animal catcher cruises the suburbs of Miami in a sport-utility vehicle capturing stray lions, tigers, cougars, rheas, macaques—even, once, a bison on the freeway. The man's business card shows a photograph in which he and three friends hold up a twenty-two-foot Indonesian python they extracted from a burrow beneath a suburban Miami home.

The invaders are legion: escaped pets; sport fish and garden plants run amok; bugs that came hidden in the foliage of introduced garden plants; pests that were introduced to control other pests, with greater or, usually, lesser success. The African clawed frog, an adaptable and omnivorous amphibian, was imported in the 1940s and '50s for use in diagnostic pregnancy tests. (When injected with the urine of a pregnant woman, the frog releases eggs—the telltale sign.) The animal's own reproductive habits were not carefully monitored, however, and by 1969 it had established wild populations in California, where it eats young trout. The invaders come as seeds, as spores, as larvae; they are four-legged ungulates set loose to roam. They come in crates, on crates, in cargo containers, and in the ballast water that ships carry to counterbalance the weight of cargo containers. Fish have spread with the openings of canals; plants have spread along railbeds; sponges have spread on the bottoms of boats. Tens of thousands of species—most of the world's fauna, minus the insects—can be and are legally imported into the United States through the mail. In recent years U.S. health officials have grown concerned at the spread of the Asian tiger mosquito, which carries dengue fever in its native continent and arrived in Houston in the mid-1980s. Laying its eggs in the rainwater that collects in used automobile tires, the insect has spread with the used-tire trade—a billion-dollar-a-year industry that sends used tires from Asia to the United States to be shredded, recycled, and reconstituted into newer used tires—to more than a dozen states and the Caribbean. It is the quintessential traveler, migrating on the wheels of yesterday's travel, which have taken on a migratory life of their own.

The invaders are from anywhere, going everywhere. The international newsletter *Aliens,* to which I briefly subscribed, provides updates on Indian house crows in Zanzibar, Argentine ants in New Zealand, North American crayfish in England, and the northern Pacific sea star in Tasmania. Australia—which for years has suffered the terrestrial deprivations of introduced rabbits, dogs, cats, camels, and ravenous, poisonous cane toads—has lately begun to focus on the intruders in its waters. The more prominent additions include the European green crab, a marauding crustacean that threatens the nation's nascent shellfish industry, and a host of toxic single-celled plankton that, when eaten by people who eat shellfish that have eaten the plankton, can prompt an unpleasant, sometimes fatal respiratory attack. Italy is battling the insurgence of the American gray squirrel, which has replaced the native red squirrel in much of Europe. (The city of Moscow is so lacking in red squirrels that it has established a special breeding program for them in a local park.) In Antarctica, ostensibly the most remote continent on Earth, researchers recently found that emperor penguins had been exposed to infectious bursal disease virus, a pathogen that normally affects domesticated poultry and is thought to have reached the penguins through garbage thrown overboard from passing ships. Feral goats in the Galápagos have caused so much erosion to the summit of Isabela Island that they have altered its rainfall pattern and water cycle; they have, in effect, changed the local climate. Even Darwin's finches are at risk; scientists recently found that finch nests in the Galápagos are infested with the larvae of an exotic parasitic fly. At night the maggots emerge and, vampirelike, suck the blood of nestlings, killing as many as one in six.

How to quantify the impact of alien species? How to grasp their toll? Wielding a financial yardstick, the federal government has estimated that between 1906 and 1991, seventy-nine nonindigenous species—including, most notably, the European gypsy moth and the Mediterranean fruit fly—had cost the nation ninety-seven billion dollars in damages, or about a billion dollars a year. Recently a group of Cornell University researchers raised the estimated damage report by two orders of magnitude, to 138 billion dollars a year. The South American fire ant costs Texas half a billion dollars annually in damage and control costs. Cleanup and control of the zebra mussel: five billion dollars annually. The Russian wheat aphid: 173 million dollars. Parasitic lamprey eel in the Great Lakes: ten million dollars. Introduced diseases of lawns, gardens, and golf courses: two billion dollars. Shipworms: 200 million dollars. Brown tree snake control and research: six million dollars. As for "emerging diseases" like West Nile virus, their cost is typically counted in human lives lost—by which measure, at least in Western nations, one is considered too many.

Ecologists work a different calculus. With exotic plants and organisms moving more readily from place to place, an increasing number of native species—residents of the planet's backwaters, unable to cohabit with zoology's rising cosmopolitan class—are being pushed out of existence. In 1991 the U.S. Fish and Wildlife Service estimated that one hundred and sixty species officially listed as threatened or endangered owe their status, at least in part, to competition with or predation by nonindigenous species. The more recent Cornell report estimated that more than four hundred species, nearly half the species on the endangered species list, are at risk. The Harvard biologist Edward O. Wilson has claimed that the introduction of alien species is second only to habitat destruction as the leading cause of extinctions worldwide.

Already alien species are so prevalent in the public eye that they meld into quotidian experience. They appear so regularly in news articles that it is sometimes difficult to grasp the entirety of the phenomenon—to see the forest, as it were, for the alien trees that keep sprouting up in it. Possums in New Zealand. Pine trees in Africa. Giant hogweed in Slovakia. Nature is entering a new era—the Homogecene, one scientist calls it—wherein the greatest threat to biological diversity is no longer just bulldozers or pesticides but, in a sense, nature itself. A creeping sameness threatens, wrought as alien species insinuate themselves into the Darwinian fabric and gradually, almost imperceptibly, supplant it. One biologist I met expressed the cost as a personal one. "There's a loss of the features that allow you to describe where you live. When you characterize where you're from, you look out your window at the plants and animals, even if you don't notice them immediately. I think there's something terribly wrong with the loss of a distinct sense of being. It comes down to: Where is my home?"

It was by way of fathoming that riddle, the nature of one's home, that I decided for a time to quit my own. Those two words, *nature* and *home*, do not customarily share the same sentence. Nature is understood to be a separate realm, wild and unsullied by the imprint of human hands and feet; as the latter multiply, so the former is diminished, to the point where nature threatens to disappear entirely, if it has not done so already. As I continued reading about alien species, however, I began to wonder if perhaps a more nuanced outlook was required. After all, nature has been moving around for hundreds of millions, even billions of years; that is the essence of biogeography. If anything, today's nature seemed to having a field day, albeit at its own expense. Was this a new kind of nature, or the old kind gone amok? What, in this rapidly changing world, *is* nature?

I realized too that, as a city-dweller, perhaps I was poorly situated to address such questions. Cities are by people, for people; that is why I enjoy living in one. What counts as nature here would disappoint a purist. The rambling wildness of Central Park in fact was wholly cultivated by a landscape architect; the polar bear in the park zoo saw a psychiatrist for a while to cure its compulsive pacing. Once, standing on the back terrace of my apartment (my rent-stabilized indoors came with a rent-stabilized square of outdoors), I saw a neighbor across the courtyard waving her hands to catch my attention. Her pet African gray parrot had flown off. Had I seen it? I remember thinking, in a very unneighborly spirit, *Nevermore!* and feeling slightly pleased that the city had gained a touch of African grayness, if only to counter the ubiquitous grayness of pigeons. The world of us was too much with me. To truly grasp the threat posed by ecological invasion, I would need to sweep aside the homogenizing scrim of humanity and seek out an unfiltered nature, heterogeneous and raw. I would find the border where people end and nature begins and would boldly step across it.

So I set out. What started as a mild interest in tracking the course of cross-species migrations quickly blossomed into a raging obsession with all things weedy and unwanted, and it soon had me tracing paths as global and intersecting as the species I was after. I attended conferences devoted to introduced species, nonindigenous species, aquatic-nuisance species, marine bioinvaders; I collected the written minutes of alien-species symposia held in Norway; I ordered illustrated binders describing every exotic aquatic plant and animal in Tasmania. My search for the brown tree snake was only the beginning. I crossed seas by ship; I went on the road; I crawled through the darkened underworld and met a ghost named Polyphemus. All I lacked was a white whale. Thoreau once suggested that a traveler need never leave his home: he could forgo South Sea expeditions and explore instead the inner world of being. Forget that. I would find my way to the end of nature, or it would be the end of me.

In the Serpent's Embrace

1

The airport shuttle driver in Honolulu laughed when I told him where I was headed. "Watch out for snakes," he said.

Nine miles wide, thirty miles long, and a seven-hour flight across the date line from Hawaii, Guam is, for the moment, one of the world's more unusual cultural cauldrons. It is the most southern member of the Marianas archipelago, a five-hundred-mile string of islands that erupted from the ocean floor four million years ago and which, were it not for seven vertical miles of seawater, would comprise the highest mountain chain on Earth. The first inhabitants were the Chamorro, a group of South Asian lineage that flourished from thirty-five hundred years ago until shortly after the sixteenth century, with the simultaneous arrival of Spaniards, Catholicism, influenza, and smallpox. Today their descendants on Guam mostly occupy the southern third of the island, speaking a mixture of Spanish and ancient Chamorro, cruising through sleepy, palm-lined villages in lowrider pickup trucks, throwing Sunday fiestas for their patron saints, all in all presenting the appearance of East L.A. on an extended tropical vacation. In 1898 the United States acquired Guam, as well as Puerto Rico and the Philippines, from Spain under the Treaty of Paris. This tenure has been interrupted only once, on December 8, 1941—Pearl Harbor Day behind the date line—when the Japanese invaded from their outpost on the neighboring island of Saipan. The United States regained custody thirty-one months later, in one of the bloodiest battles of the Pacific. Since 1944, the northern third of Guam, a jungly limestone plateau, has been the almost exclusive purview of the U.S. military, which maintains a major air force base there and, until a decade ago, several hundred nuclear warheads. Complaints are muted: the military and the Guam government, known as GovGuam, together employ more than half the island's 150,000 residents. Officially, Guam is a U.S. territory, a privilege manifest in one nonvoting congressman and two tourist slogans: "The Gateway to Micronesia" and "Where America's Day Begins."

And what a beginning. The northern and southern thirds of Guam join in a narrow isthmus of common ground: strip malls, fast-food joints, high-rise resorts, an international airport, and a sleepless snarl of traffic. At what stoplight does the district of Tumon become entangled with Tamuning? Where does Tamuning let off and Agana begin? For most travelers from the U.S. mainland Guam is a tiresome stopover on the way to somewhere else: Palau, Pohnpei, somewhere quieter, more paradisiacal. For international travelers however—particularly the burgeoning number of Japanese honeymooners, businessmen, vacationers, and bargain shoppers—Guam has become an end in itself. You can, for an immodest fee, discharge your choice of firearm (illegal in Japan) in any of several local shooting galleries, or dress in cowboy attire and ride a pony around a small ring in a back lot. You can shop tax free at the world's largest KMart, provided you can find a parking space: the lot is filled to capacity every night until the nine o'clock closing. You can stay in one of the many hotels that cater exclusively to Japanese guests. In 1972, Shoichi Yokoi, a sergeant in the Japanese army, emerged from the remote cave he had hidden in for the past twenty-eight years, unaware that the war had ended, and no doubt perplexed, on seeing so many Japanese billboards, as to which side had lost.

Into a world thus made, the brown tree snake somehow found itself. One local theory maintains that the American military set the snakes loose to get rid of the rats. No, another resident says: the snake swam from the Philippines, fifteen hundred miles away. Most biologists champion a cargo-based explanation: *Boiga irregularis* arrived some time around 1949 from a military base in the Admiralty Islands, near New Guinea, coiled in the dashboard of a jeep or in some other bit of wartime salvage. Until then, Guam had no proper snakes to speak of. What a novelty, then, to see one. SEVEN FOOT SNAKE SLAIN HERE, one headline exclaimed. Another announced: NAV MAG MEN FIND LARGEST SNAKE TO DATE—8.5 FT. LONG. And simply: SNAKES ALIVE! "Because they eat small pests and are not dangerous to man," the *Guam Daily News* reported in October 1965 alongside a photo of a seven-foot snake killed at the United Seamen's Service Center, "they may be considered beneficial to the island." Several equally large specimens were remarked upon in the local press in ensuing years, including a snake that one Mrs. Edith Smith found slipping across her neck at four o'clock one morning in June 1966.

At first, sightings were limited to the area around the military seaport at Apra Harbor, several miles south of Agana. By 1970 the snake was making appearances island-wide. Other animals began disappearing: not just "small pests"—in a 1989 poll, more than half the respondents attested that the number of rats around their homes was definitely decreasing—but also chickens and eggs, domestic pigeons, guinea fowl, ducks, quail, geese, pet parakeets and finches, piglets, and cats. One day a chicken is in the cage; the next day, there is a snake too fat to escape. Gradually the snake began attempting larger prey. Between September 1989 and September 1991 the Guam Memorial Hospital recorded seventy-nine bites by brown tree snakes. Of those victims, sixty-three—80 percent—had been sleeping in their homes at the time; and of those, half were children under age five, including two infants who were bitten while sleeping between their parents. Yvonne Matson's experience typified a mother's terror. Early one morning she awoke to the screams of her infant son; racing to his room, she saw a five-foot brown tree snake entwining him from leg to neck and gnawing on his left hand. Months later, after her mother was bitten, Matson hired an exorcist.

As fast as the snake spread, word of it spread faster, to Hawaii, California, New York. A *Wall Street Journal* article claimed that the snakes "hang from trees like fat brown strands of cooked spaghetti." But I never saw that. None crawled in through my hotel sink. None lurked in the grass island of the Taco Bell parking lot. I drove around for two days after my arrival, asking after the snake.

"I think I saw one a couple of years ago on the road when I was driving, but it was night."

"My friend saw one a couple years ago when he was driving."

"I seen more snakes in Texas!"

A farmer described seeing a snake burrowing into the nose of his pet goat, which died a day later. "And you know, the color isn't brown. It's blue!"

In short, a dissonance has arisen on Guam, a discrepancy between the heard and the seen—a gap of uncertain size, yet wide enough to rankle. One evening I attended a cocktail gathering for guests of the Guam Hilton. The general manager, an amiable German transplant, assured me that he had encountered only one snake—"may it rest in peace"—on the hotel grounds during his six years there. He sounded sincere, so I was surprised a few weeks later to read an article in the *Los Angeles Times* in which an anonymous Hilton groundskeeper confessed to decapitating twenty snakes a month with his machete. I called to confirm. "I'm not at liberty to discuss that," said the hotel employee I'd been directed to. The line promptly went dead. Calling back, I was connected to a supervisor. He shouted into the phone: "Is everybody trying to sink the tourist industry? Everybody's got this twitch about the brown tree snake and how it's devouring Guam! I've lived here for three and a half years, and I've only seen two snakes. One was in a bottle—somebody caught it—and the other was squashed on the road. I tell you, I live in the boonies, and if I thought that brown tree snakes were dripping from the trees like spaghetti, I wouldn't have my children running around outside. All you people are making a mountain out of a molehill!"

So I'd come looking for a snake, a snake that clearly had no intention of showing itself anytime soon—indeed, a snake that had successfully navigated eons of evolution precisely by going about its business unseen.

"The snakes are extremely averse to light," Earl Campbell explained to me at his field site one evening. "During the day they take cover in places where people aren't likely to encounter them. Many of my friends have lived on Guam for years and never seen snakes."

Campbell, a young herpetologist from Ohio State University, suggested I might have some luck in his company. He worked under the auspices of the Brown Tree Snake Research Program, an array of federally financed research projects established in 1988 that aims to eliminate the brown tree snake population—or, more reasonably, to prevent its spread to Hawaii or anywhere else beyond Guam. Campbell's interest was snake barriers: real barriers, actual physical fences that would confine the snake to certain areas or, conversely, keep it out of other areas; and also figurative fences, general strategies of corralling the animal, reducing its numbers, slowing it down—barriers of human intelligence unsurmountable by any brown tree snake. To halt the enemy, however, one must first know something about how it operates; and until the brown tree snake appeared on Guam in profusion, virtually nothing was known about its basic biology. To learn that, a researcher first must find a brown tree snake.

Athletic and prematurely graying, Campbell wore that grizzled look endemic to graduate students everywhere. His research site sat at the remote northern tip of Guam, on Northwest Field, a forested limestone plateau above the sea. During the Second World War the area held numerous military barracks, basketball courts, runways, and ammunition dumps. Afterward, it devolved into a scruffy jungle of tangantangan trees—a gnarled, fast-growing Central American species that was planted en masse after the war to prevent erosion and has since come to provide abundant habitat for brown tree snakes. The snake is largely nocturnal; Campbell had become so too. One evening, well past dark, he arrived at the site for a few hours of research dressed in proper fieldwork attire: T-shirt, shorts, and sandals. A crescent moon hung low above the tree line; the humid tropical air resonated with the chirring of locusts. Campbell donned a helmet with a headlamp attached to it, flicked that on, then headed down a path made narrow and low by dense foliage and dangling vines.

"Looking for a snake in the forest is sort of like being in a demented *Where's Waldo?* book," he said. He crept forward, turning his head from side to side, illuminating knotted boughs with the ray from his headlamp. "My average here is one an hour." He conceded that this number did not sound impressive. Nevertheless, he said, this patch of forest contained the highest concentration of brown tree snakes anywhere in the world. Some months earlier, Campbell had traveled with two colleagues to Australia to study the brown tree snake in its native range. In eighteen nights of looking, they together found just three snakes: one brown tree snake and two of another species. "It's incredible how well the brown tree snake does on Guam," Campbell said with a hint of admiration. Exactly how well it does, and why, are among the many things he and his colleagues would like to identify. This night's mission was part of what Campbell described as his "mark-recapture" work. Any brown tree snake he nabbed would be weighed, measured, tagged for identification and then set free, perhaps, he hoped, to be captured again days or weeks or months later. The cumulative data would shed light on the growth and movement habits of the snakes over time. "It turns out these snakes really migrate a lot," he said. "It's only when you get a barrier that you stop that movement."

Over the years, the proliferation of exotic species around the world has spawned its own field of scientific inquiry, known to its practitioners as invasion biology. The field marks its formal beginning with the 1958 publication of *The Ecology of Invasions by Animals and Plants*, by the English ecologist Charles Elton. Individual cases of biological introductions and invasions had been remarked upon in scientific journals for some time before that, but Elton was first to address such incidents as part of a large-scale, advancing phenomenon: "We must make no mistake: we are seeing one of the great historical convulsions in the world's fauna and flora." Elton covered a broad swath of history and territory—from the introduction, in 1890, of starlings into the United States (the man responsible, a Brooklynite named Eugene Schiefflin, intended to release into Central Park all the birds mentioned in the plays of William Shakespeare) to the blight wrought by *Endothia parasitica,* the Asian fungus that would eventually kill 75 percent of the nation's chestnut trees. At the time, ecology itself was a relatively new phenomenon, eager to be distinguished as a modern science apart from the gentlemanly, hunt-and-collect endeavor known for generations as natural history. So Elton strove to discern underlying patterns of invasion, to forge theories about the hidden structures of ecosystems and the manner of their unraveling. His treatise, *The Ecology of Invasions,* was as much a clarion call to his colleagues as it was to the public at large. Elton wrote, "We might say, with Professor Challenger, standing on Conan Doyle's 'Lost World,' with his black beard jutting out: 'We have been privileged to be present at one of the typical decisive battles of history—the battles which have determined the fate of the world.' But how will it be decisive? Will it be a Lost World? These are questions that ecologists ought to try to answer."

Ecologists are still at it—in greater numbers than in Elton's day, with more sophisticated instruments and analytic tools at their disposal and many more case studies to select from. To a degree, invasion biology can be thought of as evolutionary biology turned inside out. As Darwin recognized, the appearance of a new species is frequently the downwind result of an invasion—a finch blown off course, say—that occurred generations earlier. Contemporary evolutionary biologists are still working out the subtleties of Darwin's insight: the competitive tensions that ripple among species; the genetic mutations and recombinations at the heart of evolution; the isolative acts eons ago that set it off. The invasion biologist, in contrast, is drawn to those first critical moments of colonization under way right now, all around: the incidents of travel, the tooth-and-claw contests that unfold in the subsequent days, weeks, months, years. Why are some organisms more successful invaders than others? Why do some ecosystems seem particularly susceptible to intrusion? Are there general laws of invasion? Is it possible to predict which ecosystems are at risk—or, better, to predict when and where the next invasion will occur? Whether a particular introduction leads to any evolution at all—whether the native organisms can survive the "stranger's craft" in sufficient numbers to adapt—is for biologists fifty or a thousand or a million years from now to determine. The invasion biologist is ecology's emergency-room physician, discovering how ecosystems work by watching how they fall apart.

In the rapidly growing annals of invasion biology, Campbell's research subject has become an iconic figure. Upon arriving in Guam, the brown tree snake entered a monopolist's heaven. All the forces that work to reduce its numbers in Australia and New Guinea—the other snakes that prey on it or vie for the same food; the mites, ticks, bacteria, and protozoa that claim its vital juices; all the predators, competitors, and pathogens; all the elbowers and energy sappers—were absent on Guam. The result was "ecological release": unhindered proliferation, an explosion of snakes. The birds of Guam were not equipped for the challenge. Left alone for millions of years on an island without predators, many species evolved a docile manner that left them defenseless when, at this late date, predators arrived. Biologists refer to this behavioral poverty as "island tameness," and it is by no means limited to Guam. One study has found that of two hundred seventeen bird extinctions in recent history, two hundred occurred on islands.

The first of Guam's birds to go extinct were the bridled white-eyes, the smallest and least suspecting of the lot. One researcher noted that when the white-eyes were kept in a large outdoor cage, they always slept wing to wing and could be plucked off one at a time with no upset to the others. The snake proceeded to eliminate larger and more wary birds: the rufous fantail and the Guam flycatcher, each about the size of a sparrow, in 1984; the Mariana fruit dove, a sort of multicolored pigeon, in 1985; and in 1986, both the white-throated ground dove and the cardinal honeyeater, which looks something like a tanager with a curved beak and a black eye mask. I know their appearances only through illustrated books. Of the thirteen forest bird species originally native to Guam, nine were extirpated from the island. Four of those nine persist elsewhere in the Mariana Islands. Three were indigenous—native to Guam and to Guam alone—and are now utterly extinct. Another two exist only in captivity: the Micronesian kingfisher, in zoos on the mainland, and the Guam rail, a ground-dwelling bird with the body of a guinea fowl and the neck of an egret. Like many island bird species, the Guam rail—the national bird of a birdless nation—is altogether too tame for the modern world. Well before the arrival of the brown tree snake, the human residents of Guam caught rails by the sackful. The snake also eliminated three seabird species native to Guam and at least three species of small lizard. The Mariana fruit bat, an enormous flying mammal that feeds during the daytime and numbered in the thousands only a few decades ago, has been reduced to a population of a few hundred on a forested bluff below the air force base. (That they are the primary ingredient in fruit-bat soup, a local human delicacy, does not help their prospects.) Guam's ornithologists are left with a scientific task distressingly similar to Campbell's, of studying subjects that are largely absent—and truly so, not merely apparently so.

Other than the Guam rail, the only native forest bird still found on the island is the Mariana crow, better known locally by its Chamorro name, the aga. It is a majestic specimen of the genus *Corvus*, upward of fifteen inches tall and coal black. In sunlight its head and back refract a dark green gloss, its tail a scintillating cobalt. Relative to continental crows, the plumage of the aga is, in ornithological terms, "lax," with a misadjusted feather tufting out here or there, conferring on it the air of a stylish, easygoing monarch. The aga's large size meant that it was among the last birds on Guam to be threatened by the brown tree snake. Nonetheless, by 1985 the forests of Guam contained only a hundred individual agas. Ten years later only twenty-six could be accounted for. By early 1997 that number had fallen to fourteen: four mated pairs and six unpaired "singletons." A few hundred agas—a sizable if not exactly vibrant community—persist on the nearby island of Rota, which is still free of brown tree snakes. Rota is the lone remaining wilderness harbor of the aga. The only other population is a small cluster of political exiles: ten crows spirited from Rota some years earlier and sent to breeding programs in Texas and Virginia.

The continuing survival of the agas on Guam today depends upon an elaborate operation run by Celestino Aguon, a biologist with Guam's Division of Aquatic and Wildlife Resources. By night I sought snakes with Campbell; by day I followed Aguon and the birds. Lately Aguon had taken a keen interest in one pair of agas in particular, an older couple known as 11B6. The pair had chosen to build their nest—a large, intricate platform of sticks, vines, and rootlets—at the top of a tree in a forested corner of the air force base, just off a long avenue of concrete bunkers that in years past had been used to store ammunition. The tree that held the nest had itself been transformed into something akin to a military fortification. The trunk, for several feet off the ground, was wrapped in a skein of wires, and the branches just above it had been cut short to prevent snakes from creeping onto them from the surrounding trees. A solar-powered battery, the size of a small toolbox, sat on the ground nearby and fed the wires with a steady current of five thousand volts. Through various experiments, Aguon had found that this arrangement was sufficient to dissuade snakes on the ground from climbing the trunk. I followed him to a blind—a square roof of black plastic stretched between branches—about a hundred yards from the fortified tree. From here I could see the female half of 11B6, her head black and bright, peering over the edge of the nest. The couple looked trapped in its safety, like those people in Manhattan who secure their apartments with eight locks on the front door, but of course the crows, with no roof above them, were free to fly off at any moment. Indeed, crows are intelligent and fiercely observant, and any prospect of excessive harassment—by a roving male aga, hovering military helicopters, or too many researchers lingering too near for too long—can spur a pair to abandon their nest and start again elsewhere, requiring the human encampment to track down the nest and impose its siege of vigilance once again.

An organism introduced to a new environment must overcome several natural barriers. Climate and nourishment, principally. Can it take the weather? Can it find enough to eat? These are immediate challenges, and they help to explain the brown tree snake's success on Guam, just as they help to explain why many other introduced organisms fail to take hold, or why some biological invaders make only limited progress. A less immediate but no less critical barrier to invasion is reproduction. To prosper, an invader by definition must procreate. Even if, like the snake, it can arrive laden with fertile eggs, eventually it or its offspring must find a mate. The odds of doing so depend on many variables, including the ratio of males to females, the speed with which individual newcomers disperse into the new territory, and the size of the territory they are entering. It boils down to sex: Can male and female find each other, and can subsequent generations do so with sufficient frequency and reproductive success to create a viable population? There is a numerical threshold above which a species can outlast extinction and below which it cannot; that number varies widely among species. This threshold applies not only to invaders but to native species as well; it governs the struggle to occupy old ground as well as new. The agas remaining on

Guam were approaching this barrier in a descent from above. They are victims of what biologists call "sexual disharmony," the scientific way of saying they may be doomed to die alone.

Nor, as Aguon soon explained, is that the agas' only problem. In 1990, when it became clear that the agas were losing their eggs to the snake, nest protection began in earnest. Aguon designed and erected the electrical barrier and set several snake traps nearby. The first night out, he caught seven brown tree snakes near the nest. Within a week he had trapped twenty-one outside the electrical barrier. In three weeks he caught thirty. The nests, although still beacons for snakes, were now demonstrably safe. The following year the protected nests produced a couple of hatchlings, including the future Mrs. 11B6. In unprotected nests, female agas were known to lay one or two eggs in a clutch. Now Aguon began to document clutches with three eggs. In addition, he discovered that if the eggs were removed from the nest soon after laying and placed in a laboratory incubator, he could induce the female to lay again, thereby increasing the number of potential aga hatchlings. That was the good news. The bad news was that of the forty-one eggs laid in various nests between 1992 and 1995, only four hatched. Only six agas have fledged since 1989. There are many reasons why an egg might fail to hatch—cracks, inadequate incubation in the nest, calcium deficiencies—but the main one in this case, as Aguon discovered with chagrin, was that most of them were not fertile. In the time it had taken him to make nests safe for upcoming agas, the remaining adults had aged irreparably. "We couldn't keep up the recruitment of younger crows," Aguon said. "We solved the protection problem, but now we're facing infertility." He motioned to the nest of 11B6. "This particular pair hasn't produced anything in years. Their pattern is to build a nest, abandon it, build, abandon. My conclusion is they're beyond the age of reproduction. They're doing everything you'd see at a normal nest. I don't know what's in it. I'll give it a few days and see."

Slight of stature, with a full beard and tired brown eyes, Aguon—"Tino" to his friends—is himself something of a rarity, the only native Guamanian biologist in the history of Guam's wildlife agency. As the number of agas on Guam has dwindled, so have the hours he allots for sleep: his is often a schedule of eighteen-hour days and seven-day weeks. Colleagues speak of him with a mixture of admiration and pity.

"Tino has the hardest nut to crack. He was involved in projects with three other species that have disappeared. Our joke is, whatever Tino touches goes extinct."

"He's trying to handle the last bird—the last egg of the last bird—on Guam. That's a tremendous amount of pressure. And there are no new Tinos coming along."

Island species, isolated from predators and the vigors of competition, often thrive; yet the same conditions may also render them vulnerable when their environment undergoes rapid change. The same could be said of island biologists. "The people on Guam have burned themselves to a frazzle," said a former colleague of Aguon's. "They are the hardest-working people I know. But there's a sense of panic. They've seen more taxa go extinct under their watch than anyone else. It's not easy: that might be the last crow out there if you don't find it. That—and the isolation. You don't have the intellectual exchange that you do in Hawaii. Ideas happen somewhat in a vacuum. You feel a little alone, a little betrayed."

By 1995, only a single pair of agas was producing properly fertilized eggs. That season, before the male disappeared for good, they laid two. Twice Aguon climbed the tree, lifted an egg from the nest, tucked it into a small box bedded with sawdust, put that in a larger, temperature-regulated container, and transported it back to the incubator in the lab. The first egg, bearing a male, hatched in captivity on April 2, 1996; the second, bearing his sister, hatched two days later. These are the last two native agas to have fledged on Guam. The female is named Nancy. When I last saw her, she was perched in a large cage in a forest grove not far from the abandoned ammunition bunkers. Before a juvenile captive-bred crow can be released into the wild, she must be "hacked out": set outdoors in a cage to acclimate, her door eventually opened, and her food gradually decreased from week to week until she is forced to venture out and forage on her own. Aguon drove the back roads to Nancy's cage, parked several dozen yards away, and shut off the engine. Nancy was an elegant youth, already too large to be taken by a snake. On sighting Aguon, she cocked her head and began flapping her wings, the numerous marker bands on her legs jangling like bracelets. She was a girl on the lookout, sharpening up for the crow equivalent of Saturday night, too young yet to sense the coming days of heartbreak. Aguon said, "The unfortunate thing is, we're unable to get Nancy mated up."

He backed out of Nancy's grove and wound his way past the ammunition bunkers, out the main gates of the air force base, down the main road past concrete suburbs, and into the fray of stoplights and construction detours, toward his office. Aguon has little spare time these days in which to indulge his exhaustion; with the time he does have, he often visits local classrooms to work with human fledglings. "If we're gonna succeed in conservation efforts, kids are the place we've got to start," he said. "I don't really stand in front of them and give them a doom-and-gloom scenario; they don't have much feel for numbers. I give them a little taste of biology and geology and connect it with the animals here. I always take a branch and let a child hold on to it. 'This is a branch. Where does it come from?' A *tree.* 'What do you call a whole bunch of trees?' A *forest.* 'What sorts of things live in the forest?' I don't just tell them that this bat is an endangered species. I ask, 'Where does it live? Where does it get food?' I keep going back to the forest, so they know not only that the snake is a problem, but that losing the forest is a problem for some of these animals. It's kind of fun. The kids have taken me out of my element a little. You can hold their attention for a whole hour—it's unbelievable." Nevertheless, he said, for some children the learning curve is dismayingly steep. "They're growing up in an environment devoid of wildlife. When I ask what lives in the forest, some say, 'Crocodiles.' It's something they've seen on television."

2

By ten o'clock at night the heat of the day had dissolved, and a feathery dew began to swirl through the trees of Earl Campbell's research site at Northwest Field. From deep among the branches, small amber jewels—the reflective eyes of nocturnal moths—glittered back at the light on Campbell's helmet. The eyes of the brown tree snake, unlike those of most snakes, would not reflect the light from our headlamps, rendering the animal doubly hard to detect. Every few steps brought an entanglement of spiderwebs—gummy wisps, woven on a scale to rival tents, that glued themselves to hands and face. The effect was sepulchral, and it had been even more so earlier in the day, in full sunlight. No sound but for passing wasps and the drone of an overhead plane—an enveloping aural emptiness created, I realized, by the absence of any birdsong. As for bird sightings, I could count them on one hand, even after several days: one drongo, a shrike from Southeast Asia; a species of turtledove that had immigrated from the Philippines several years ago; and a few bedraggled Eurasian sparrows that stuck close to my hotel. "It's amazing, just hearing the wind blow through the forest," Campbell said. "It's warm, it's lush, it's tropical, but there's no noise."

The snake owes its success on Guam as much to its own remarkable biology as to the pleasant climate and abundant food that greeted its arrival. Although the brown tree snake has been known to grow longer than ten feet, most are yardstick length and thin as a human pinkie. It is also a master climber, endowed with a slender prehensile tail and strong enough to hold three-quarters of its body upright as it stretches for a distant bough. (When a researcher attempts to pull one from a tree, the snake can tie itself to the branch with a half-hitch knot.) The sum product is a supremely cryptic creature: nocturnal, capable of sliding soundlessly from branch to branch, yet able in an instant to strike a pigeon from midair. The rear of its mouth holds several small sharp fangs that can grip prey until its mild venom asserts a hold. With its flexible jaw, the snake can devour creatures more than half its own meager weight, including at least one Labrador retriever puppy.

Precisely because of this battery of traits, *Boiga irregularis* had long been considered a specialized creature, built for an arboreal existence and little more. But behavior, even in a pea-brained reptile, is surprisingly flexible. On Guam, the snake forages as easily on the ground as in the trees, and it will consume anything that smells of blood: dog food, spareribs, soiled tampons, rotting lizards, paper towels, maggot-infested rabbits, raw hamburger, even the foam plate that the hamburger comes on. What scientists thought was one serpent in effect is two: the native and the colonist, the preinvasive and the postlapsarian. The hunter became a scavenger. "This sort of thing catches the attention of a herpetologist," one herpetologist told me over the phone. "It's unheard of in snakes. I came back from Guam and tested it out, and by God it's true—they do eat dog and cat food. And they will certainly eat bloody raw meat. There is a certain mystique among breeders that a captive snake must be fed live, natural-looking prey. Now along comes this snake that you can just as easily feed sausages."

Environmental scientists sometimes characterize a species in terms of the "role" or "function" it plays in an ecological community. Charles Elton popularized the term *niche* to describe something similar—the working relationship between an organism, its food, and its enemies. "When an ecologist says, 'There goes a badger,' he should include in his thoughts some definite idea of the animal's place in the community to which it belongs, just as if he had said, 'There goes the vicar.'" The brown tree snake illustrates one hazard of this perspective. It is all too tempting to assume that an organism's "ecological role" is inflexibly inscribed in its biology. The human mind may be more niche-bound than any physical species.

Indeed, the perennial mystery—and marvel—of ecological invasion lies in the erratic link between biological behavior and geographic location: that a species can be placid and contained in one location (its "home range") yet may become a reproductively raging menace in a foreign setting. The vicar at home becomes a devil abroad. In *Ecological Imperialism*, the historian Alfred Crosby notes that in the nineteenth century, British botanists expressed amazement that so many English plants had spread to both Australia and North America, yet very few plants from either of those continents had spread to England, despite plenty of human traffic between. Joseph Hooker marveled at "the total want of reciprocity in nature" and expressed his thoughts to Darwin. "We have the apparent double anomaly, that Australia is apparently better suited to some English plants than England is, and that some English plants are better suited to Australia than those Australian plants were which have given way before English intruders."

Such anomalies have offered the tantalizing hint of an underlying pattern—of natural selection, of biogeographic direction. Biologists have avidly sought to identify the common traits of successful invaders and, with that list, to predict which species are more likely than others to successfully invade. (Agricultural companies often rely on such lists to argue that their genetically modified plants will not become weeds.) Increasingly, however, many invasion scientists have begun to question the value of these trait lists. For example, it is commonly assumed that a species that reproduces faster will make for a better invader, but the supporting evidence "is thin and contradictory," writes the biologist Mark Williams in his book *Biological Invasions*, a recent survey of invasions research. One basic problem is that the "intrinsic" rate of increase measured for a species in a laboratory setting can differ widely from the rate measured in the wild—and the wild rate can vary depending on location and circumstance. Even then, fast means nothing. Humankind, arguably the most successful invader in biological history, grows at the very slow rate of 3 percent a year. Some scientists contend that successful invaders are often "opportunists," but others wonder if the term isn't simply a post hoc explanation masquerading as a noun.

Although biologists on Guam had followed the decline of the forest birds with growing alarm since the 1960s, the brown tree snake was not considered seriously as the culprit for another two decades. The creature was easy to overlook—not visibly abundant, apparent only by the absences that accumulated around it. No previous record existed of a snake obliterating an entire island of birds, and even if such a thing were conceivable, the little evidence then available did not suggest that the brown tree snake would prove to be the historical groundbreaker. Any and every other cause seemed more likely: remnants of the DDT that had been sprayed as a defoliant during the war; or a pox, akin to the avian malaria that had begun to claim forest birds in Hawaii around that time; or the loss of habitat as Guam's forests were renovated into houses, roads, malls, hotels.

Beginning in 1982, Julie Savidge, then a graduate student working in Guam, studied the possible causes and one by one dismissed them. She took a hard second look at the snake. From conversations, interviews, and newspaper accounts of sightings, she charted the advance of the snake across the island. Her map resembled a rippling pond, the first waves emanating from the military seaport in the 1950s and, by the 1970s, reaching the northern limits of Guam. She then mapped the decline of the birds: their last sightings and dwindling numbers. The two maps, of birds and snakes, were virtually identical. Two overlapping sets of ripples, one reptile dropping silently through the center of the pond. The paper she later published, "Extinction of an Island Forest Avifauna by an Introduced Snake," is now a classic: a case study of what a snake can do when left alone on an island full of birds and, more broadly, of the deceptive tranquility with which even the most ruinous biological invasion can unfold. No wonder the Hawaiians are concerned, Earl Campbell said. By the time they know for certain whether the brown tree snake is among them, it will be too late. "How do you prove to people there's a problem if it's something they can't see?"

Campbell had yet to spot a snake that evening. We had walked more than an hour, first down one dark path, then another, stumbling over roots, stalking the shadows of tree branches, when he stopped abruptly and said quietly, "There's one." The beam from his headlamp, which had been programmatically scanning the walls of foliage to either side, now illuminated one leaf in particular, on which was perched a gecko, wide-eyed and blinking. Campbell estimated its size and jotted the figure in a pocket notebook. I was underwhelmed—it was not the snake I still hoped to see—but Campbell noted that the gecko too was an invading species and in fact was a critical factor in the brown tree snake's success on Guam. "We have all these introduced lizards," he said. "There are five species on Guam. Two date back to when the Melanesians first came to this area. The other three have come during recent shippings." They are all prolific breeders, he added, and although none live in the snake's native range, they now form a significant part of its diet, enabling it to grow and multiply in a forest virtually devoid of birdlife. "That's the major reason the snakes are doing so well." Encouraged by his academic adviser, Campbell had added gecko and skink counts to his nocturnal surveys. How many, what size, where in the canopy, what time of year. From this database, one could begin to chart the demographics of the lizard population, the abundance and quality of the snake's repast. The early results, in the scientists' opinion, are sobering.

In simple ecological models, the relationship between a predator and its prey is straightforward, Malthusian. In the classic model, there are rabbits and there are lynxes eating rabbits: the lynxes thrive, reproduce, spawn more hungry lynxes—so many that soon there are fewer and fewer rabbits, fewer lynx meals, eventually fewer lynxes. The lynx population declines, the rabbits slowly recover their numbers, and the cycle of eating and eaten, supply and starvation, begins again. The situation on Guam is altogether different. Even after the forests were emptied of birds, the snake continued to thrive. Its numbers are down: approximately twenty-four snakes per hectare, from a hundred per hectare in the late 1980s. That is a major drop, yet twenty-four snakes per hectare is still four times more dense than even the most snake-infested plot of Amazon jungle. Campbell said, "That's like having a gob versus a big gob." Now the snakes subsist on skinks and geckos. And the skinks and geckos are not disappearing. In fact, because they are no longer preyed upon by birds, they are more abundant than ever. The snakes have found a renewable resource, the gustatory equivalent of solar energy. And there is evidence to suggest that the snakes are reproducing faster too, giving birth at a younger age. Few biological invaders, once they have gained such a solid foothold in their new habitat, subsequently disappear from it entirely. Any hope that the snake would eat itself out of existence appears equally groundless.

In *On the Origin of Species,* Darwin meditated on the impact that house cats might be having on the surrounding countryside. If there were more cats, there would be fewer mice. With fewer mice burrowing into the hives of their favorite snack, the bumblebee, there would perforce be more bees buzzing about, gathering nectar, pollinating the local blossoms—most notably the blossoms of *Trifolium pratense,* the common red clover, which depends exclusively on the bumblebee to complete its reproductive cycle. In short, more cats would mean more red clover. "Plants and animals, most remote in the scale of nature, are bound together by a web of complex relations," Darwin wrote. Even a small perturbation can reverberate throughout the web; the addition of even a single species can create a cascade of secondary effects. So it has been with the brown tree snake. Its success on Guam was paved in part by the invasions that preceded it—not merely the skinks and geckos, biologists suspect, but even the trees in the forest. Guam's forests, once a rich mixture of several tree species, had been devastated by the war and subsequently reseeded with a relatively uniform assemblage of tangantangan trees. The new canopy was both less dense and visually less complex in the eyes of a snake; the nests of birds now stood out far more clearly. Indeed, the first birds to disappear were not only the smallest but also the ones that nested in tangantangans. Only the more specialized species—the kingfisher, which nested in rotten holes; the Mariana crow, which nests in high trees—presented the snake with some semblance of a challenge.

With the snake now a permanent presence, biologists like Campbell have begun charting its cascading impact. The introduced skinks and geckos are thriving. So are the introduced birds: Philippine turtledoves, which reproduce more quickly than most of Guam's native forest birds ever could have dreamed; European sparrows, flourishing amid the bustle of downtown Agana; the Southeast Asian black drongo, nesting atop the island's concrete utility poles, which are too smooth and wide for the snakes to climb. And as the forest birds have disappeared, the insects they once fed on, the gnats, butterflies, and moths, have flourished. The flutter of feathered wings has been replaced by a flurry of tiny membranous ones, in turn prompting a frenzy of spinning among the spiders. All the better for hungry, multiplying skinks and geckos. The feedback loop is complete. "The invader," the ecologist Charles Elton wrote in 1958, "is therefore working his way somehow into a complex system, rather as an immigrant might try to find a job and a house and start a new family in a new country or big city." Some organisms arrive to certain failure, defeated by the wind, the soil, the hot or cold, hunger, reproductive frustration. Others, the ecological wallflowers, arrive and quietly persist, squeaking by, rooting in, thriving just below the level of human awareness, making no apparent—or at least no immediate—impact on the system as a whole. And still others are outright boors, strolling through the doorways occupied by their reclusive friends, elbowing their way toward the buffet. Soon their friends arrive, and more friends, uncles, cousins, until a point is reached when the hosts, expecting to mingle among familiar faces, discovers with dismay that their dinner party is filled with uninvited guests. New strands weave their way into the food web, the old ones unweft—singly, unobserved, supplanted.

"This place isn't truly natural," Campbell said before quitting for the night. "Most of the plants around here are Central American. Introduced species do very well here. It's a McDonald's ecosystem, and these are real fast-food animals. If people don't watch out, we're going to have the same metropolitan animals across the globe. I see the same birds in Copenhagen as I see in Toledo and Sydney. In the St. Lawrence Seaway, forty to sixty percent of the animals are introduced. I leave Guam and go to my parents' place on Lake Erie and get my feet cut up by zebra mussels. What's wrong with cosmopolitan fauna? There may be people who would like to just get it over with, because it would make it simpler to get on with business. Concerned citizens, businessmen on Guam have said, 'Well, if we can't have the native birds here anymore, why not introduce beautiful tropical birds, parrots, because tourists will like it.'"

Consider a length of thread, silken, delicate, seemingly endless. It begins in a patch of matted grass on the side of the road, it enters the forest, and, because the thread is so fine as to be barely perceptible against the blare of tropical foliage, it promptly disappears. At the far end of the thread, in theory, is a brown tree snake. At the near end, matting the grass with his hands and knees, sweating in the noon sun, tracing the thread with a forefinger, is Craig Clark.

Clark, like Campbell, was a budding herpetologist on loan from Ohio State. In the several months the two of them worked together on Guam, they had tagged, released, and gathered data on more than three hundred brown tree snakes. In that time they learned at least one essential truth: the brown tree snake is highly mobile. It is continuously on the go, moving as far as half a mile on a given night. If it has a home range, a piece of property that it considers familiar and to which it returns with some regularity, like a night-shift worker at the end of the evening, that range is sizable enough that biologists have not yet succeeded in fixing a number to it. Every organism treads a path between conflicting instincts: safety, stillness, occlusion on the one hand; novelty, movement, exposure on the other. Motion is a road to fulfillment: nourishment, reproduction, dispersion. It also renders an organism vulnerable—to predators, to the elements, to physical damage. Travel is the driving need and the Achilles' heel. The snake is nature's rule in this respect, not an exception.

Campbell focused on the snake's mobility on a macro scale, across acres and even miles. Clark would like to understand the micro scale, the up-close details of the snake's perpetual motion. When a brown tree snake moves around in the course of an evening, where exactly, meter by meter, does it go? "Perch height, perch type, incline, distance from trunk, length of branch. I want an estimate of where they're using the canopy. I want to be able to say, 'This is where the snake is going.' Is it a random pattern? What are they looking for? How much time does it spend foraging? How much time does it spend sitting around?" His project has practical implications. Although brown tree snakes will never be entirely eliminated from Guam, biologists have found that they can be trapped out of small, select areas: caught one by one in specially designed snake traps until the local population drops to near zero. But a snake trap works only if it is placed in a location a snake is likely to venture past; and those locations are what Clark would like to identify. "What species of tree do you put the trap in? How far off the ground? One snake I tracked for one hundred and eighty meters: it took me four days. Tracking techniques have been used on rodents, but nobody else has tried to do this with an arboreal snake. And with good reason: it's too much work."

Clark's field site was a patch of forest several miles south of downtown Agana on the property of the U.S. Naval Air Base. To reach it, he had to pass through a security checkpoint, drive on past a concrete building complex—still used occasionally by U.S. Navy SEALs for jungle training—and park along the shoulder of a long and mostly empty road, in front of an unremarkable wall of foliage. The forest is second growth: the short, scrubby tangantangan trees, so prevalent on Guam, interspersed with the occasional cycad, a native palmlike tree with evolutionary roots a quarter-billion years deep. A few hundred yards away, hidden on the far side of the forest, was Apra Harbor, the port through which the brown tree snake first entered Guam. Although additional brown tree snakes may or may not be arriving still through this port, more immediately hazardous materials, such as live bombs, do come and go. Clark said, "If something goes wrong, technically I'm within the blast zone."

Clark's research methodology is the clever product of modern technology and late-night tinkering. He begins with the research subject, a live brown tree snake, retrieved from a cageful kept at the lab for such purposes. Just behind the snake's head Clark attaches the serpentine equivalent of a backpack: a miniature radio transmitter, a miniature battery, and a bobbin with two hundred meters of red thread that feeds out from the inside of the spool. Total weight, approximately five grams. To prevent the load from snagging on twigs or rocks, Clark fits a small triangular piece of plastic straw in front of it, like the cowcatcher on a locomotive. Thus laden, the snake is released at night; Clark, hiding behind a blind and wearing night-vision goggles, watches it enter the forest. The next day—today—he tracks the snake's radio signal to its hideout. Then he will capture the snake, retrieve the transmitter, quickly dissect the snake, and check its stomach contents to find out what it has eaten in the past few hours.

On this particular day he was reeling in research subject No. 18. The first seventeen snakes had been trial runs to work out the kinks in the experimental setup. Clark said, "Whenever you start a field project like this, you expect a lot of hiccups. Things you don't expect to go wrong, go wrong." For example: if the snake travels more than two hundred meters, it runs out of thread, and Clark loses the snake and some expensive miniature transmitting equipment. This has happened more than once already. For this reason, Clark tries to let no more than twenty-four hours elapse between the time of release and the time of recapture.

In the next few moments Clark would discover something else that can go wrong. He extracted a radio transceiver from the bed of his pickup truck. It was three feet across and shaped like the letter H, and it had a handgrip, like a pistol. Clark held the H horizontally above his head. He flicked a small switch on the handle, and the H emitted a low whine, more or less insistent depending on the direction Clark aimed it. He followed the signal into the forest, losing it, finding it, losing it again. "It's fairly imprecise," he said. "I do a fair amount of wandering around aimlessly." And stumbling. The footing, as in many Pacific island forests, consists mostly of small, sharp blocks of limestone karst. Clark had come out here the previous night hoping to get an early bead on No. 18. Wandering in circles, with only the light from his headlamp to stave off total creaking blackness, it was all he could do to avoid becoming completely lost within forty yards of where he'd parked.

The pitch of the radio signal increased steadily as Clark moved forward, until he stood before a large pile of limestone rubble. Last night's search had ended at the same spot. "From what I can tell, the snake is underneath there somewhere," Clark said. "There's not much I can do about it except wait, and hope the batteries outlast his patience. He may sit there for a while. The snake has a really slow metabolism. If he ever moves, I can get my radio back."

We stood there for a few moments, Clark holding the whining transceiver overhead. Finally he turned it off and made his way back through the trees to his truck. He still had two hundred meters of thread to trace. He indicated a small orange flag at the forest edge, where the snake had been released and had crawled into the brush, unspooling as it went. The grass near the flag was flattened where Clark had spent the morning taking data on the first few meters of thread. Now he sat down and resumed the work. He found the thread again, a red filament woven into the fabric of the grass, and measured out another meter. With a compass he determined the snake's direction of travel, then made a note in his notebook. He identified the predominant plants in that meter of microhabitat and noted them in his notebook. From a pocket he retrieved a spherical densiometer: a curved mirror the size of a woman's compact, with a grid pattern imposed on it. By counting the number of darkened squares of mirror, Clark could quantify the amount of daylight eclipsed by the forest canopy. He jotted some numbers in his notebook, then crawled on. The details of travel: effortless for a beast to perform, painstaking for man to recover.

Two hundred meters of thread, five minutes per meter: it would take Clark all week to reach the end. If the thread moved up a trunk, into a tree, if it entangled the fruit of the thinnest upper limbs—for even the largest brown tree snakes can navigate the top of the canopy, so adept are they at distributing their weight—Clark would follow, on his stepladder. "If I thought like a snake, this would be easy," he said.

I picture him there still: crawling through the tall grass, eyes on the thread, thread between forefinger and thumb, taking the measure of weft and warp. He is inching forward, spinning in, deweaving, insinuating himself into the tapestry, into the forest, into the mind of the serpent.

3

The nerve center of brown tree snake biology rests on a neck of land at the base of the cliffs of Ritidian Point, in a low-lying, typhoon-proof concrete box. Here is the field station for the U.S. Brown Tree Snake Research Program. It is a windowless building, its corridors gloomy, labyrinthine, claustrophobic. For many years it belonged to the American military and served as the primary naval listening post for the western Pacific. It was, in effect, a giant invisible ear. Outside the building, foreign ships and submarines crisscrossed the oceanic expanse, whispering and conspiring in code; inside, within the most isolated chambers of the ear, American intelligence officers combed the waters for signals—intercepting, decoding, interpreting. Or so it is said. What actually went on was secret and largely remains so to the building's present-day occupants, herpetology grad students who wander its warren of offices and clack away on computer keyboards. Gad Perry, their supervisor, showed me down the main hallway to a pair of solid steel doors, which opened into a cavernous empty room—the former nexus, allegedly, of the naval monitoring operation. At the far end of that room was a smaller steel door, leading into a small room. It was in here, sealed and soundproofed from the world, that the real spy work went on, where the intelligence homunculi intercepted foreign messages and slipped them to compatriots in the outer room through a slot in the wall. In this inner room was yet a third door. It too was solid steel, with a circular crank handle, like a bank vault. The combination to the lock has long been forgotten, the door never opened, the secrets of the innermost room never exposed to light.

"We have no idea what's in there," Perry said.

Perry—ursine, with bright eyes above a bristly brown beard—is the on-site manager of the field station. At the moment, he was working on his laptop in one of the windowless rooms off the main corridor, checking his e-mail through a fuzzy phone line. He sat on a rickety swivel chair, his feet propped up on a second one. Gordon Rodda sat to his left, similarly occupied. In the private opinion of several of his colleagues, Rodda is a herpetologist of Einsteinian caliber. He began his career with a doctorate in animal behavior—specifically, the navigational skills of alligators in Florida. Later, on a fellowship from the Smithsonian Institution, he became immersed in the sex lives of green iguanas in Venezuela. In 1987 he was hired by the federal Brown Tree Snake Research Program and now designs and runs its science agenda. He has spent more than a decade roaming the territory, actual and potential, of *Boiga irregularis.* At one point he spent several months studying the brown tree snake in Papua New Guinea and the Solomon Islands—living in thatch huts, battling internal parasites, and sneaking into the forest at night to conduct snake surveys while trying to evade curious villagers who kept wanting to follow him. He is loose-limbed and rangy—an affable, energetic shipwreck. A red beard tends toward the untamed; on the top of his head is a small clearing, a bald spot surrounded by an unruly thicket of red hair. The root bed below, by nearly all accounts, contains the neuronal axis of brown tree snake biology. Rodda's graduate students and postdocs—Earl Campbell, Craig Clark, Gad Perry among them—pass through Guam for varying lengths of stay; they are autonomous extensions of the same neural cluster. Rodda himself is based in Fort Collins, Colorado; administrative duties keep him from traveling to Guam more than three or four times a year, for two to four weeks at a stretch. The rest of the year, it falls to Perry to oversee the myriad research projects concerned with brown tree snake control.

Control is the operative word, because no scientist in history has yet been successful in eradicating an entire population of snakes. Nevertheless, there is no lack of biologists eager to make an exceptional example of Guam's brown tree snakes. Snake poisons. Snake diseases. Snake birth control. I spoke to one zoologist with an elaborate scheme to insert into the DNA of the snake a chemical or protozoan or virus that would effectively disrupt its reproductive cycle. It is his idea that the disruptive agent could be carried by some intermediary—a Trojan mosquito of sorts—set loose on Guam to make the delivery. ("It would have to be fail-safe," he assured me on the phone. "You wouldn't want something that bites humans.") To be successful, such schemes needn't eliminate every single snake. They need only make a reasonable dent in the population: to make it more difficult for brown tree snakes to find one another and reproduce; to remove them from small, select areas of forest where biologists hope to revitalize what remains of Guam's bird population; to make sure there aren't quite so many brown tree snakes in the trees around Guam's airport, waiting to wander into the next outbound plane. Some of those projects have gained a boost from research that Rodda has conducted into the basic biology of the snake. By and large, however, the Brown Tree Snake Research Program keeps to its own specific mandate: preventing the snake from going anywhere beyond where it already is. The brown tree snake was born to move. The Brown Tree Snake Research Program was born to decipher, infiltrate, and arrest that movement.

"A lot of people on the mainland perceive the snake problem as something that's happening far, far away," Perry said. "It's been hard to convince them that it is their problem. And it is: the snake has gotten to Texas once. They could do well in quite a few parts of the mainland. South Florida—we're quite sure the snake could get established there. It may well be the next stopping place. The ecology is great: introduced rats; agricultural areas close to urban ones; introduced anoles, they'd be the primary source of food. Even downtown San Diego: nice climate, lots of sprinklers. The snake could eat alligator lizards until the sun goes down." He corrected himself: "*After* the sun goes down." If some of the questions Rodda and his colleagues are pursuing appear at first glance to be tangential or arcane, give them time, Perry said. "We're trying to be three to five years ahead of the need. People think we're working on stuff that's not relevant. But we're working on things you don't know you need yet."

Rodda and Perry have assumed an almost epidemiological outlook. The brown tree snake represents a kind of contagion, surreptitious yet virulent. It is spread not by mosquitoes, like malaria or the West Nile virus; nor, like the bubonic plague, is it spread by a combination of rats and fleas. Rather, it is transmitted by the everyday vectors of human commerce: planes, ships, vehicles of cargo. As far as anyone knows with certainty, Guam is the only place with an established population of brown tree snakes; it is the only source of the problem, the only host of the disease. Rodda and Perry would like to keep it that way; otherwise every new victim will itself become a host, an infected node, a potential source of further contagion. "The place we're trying to protect is Oahu," Rodda said. "It's the trade nexus for virtually every island in the Pacific: Nauru, Micronesia, the Marshalls, the tropical Far East. We see it as a first line of defense for Tonga, American Samoa, and all the other places that the U.S. doesn't care about."

Already the snake has appeared on several Pacific islands that conduct trade with Guam. As early as 1979, one was seen crawling from the landing gear of a military cargo plane on Kwajalein Atoll, in the Marshall Islands. In 1994 an inspector on the Micronesian island of Pohnpei found a brown tree snake crushed between two cargo containers being unloaded from a ship from Guam; the ship had already stopped at three other nearby islands. Since 1986, Saipan—the capital of the Commonwealth of Northern Mariana Islands and the weekly recipient of at least twenty cargo containers shipped from Guam—has recorded three dozen sightings of brown tree snakes; whether the snake is an established resident or only an occasional migrant is unclear. At the fish-and-wildlife office, on Saipan's main strip, a map of the island marks the location of each snake sighting with a red pin. Most of the pins are clustered around the seaport, a paved acreage of wooden pallets and idle cargo containers. Each pin is matched by an entry in the agency logbook:

> 14 May 1990. A brown tree snake was discovered dead at the airport inside an air cargo container. Subsequently, an airline employee on Guam was overheard to claim responsibility for deliberately placing the snake in the container as a joke, knowing it would cause a stir on Saipan.
>
> 4 January 1993. A visitor from Guam, who was aware of the desire to keep snakes off Saipan, reported trying to run over a snake as it passed in front of his car on the beach road at 3 a.m.
>
> March 1994. A juvenile brown tree snake was captured by an airline employee after it was spotted climbing on a cyclone fence midway between the main terminal and the commuter terminal at daybreak.
>
> 11 November 1995. A juvenile brown tree snake was observed swimming in the ocean next to a tug recently arrived from Guam; presumably the snake had just fallen off the boat.

In November 1995 four brown tree snakes were found outside a used-appliance store on Tinian, a small island forty miles north of Guam and a ten-minute puddle jump from neighboring Saipan. Tinian is historically notable for its flatness. During the Second World War, after vicious fighting with the Japanese, the Americans seized Tinian and converted it into airfields from which, on August 6, 1945, the bomber *Enola Gay* set off for Hiroshima with the world's first atomic payload. The runways are cracked and overgrown now, but with a map one day I found my way through the green labyrinth to a somber pair of plaques on the tarmac where Fat Man and Little Boy once awaited their final journeys. The rest of the island is equally, blissfully, forgotten. The sole endemic bird is a rufous finch called the Tinian monarch. The island's four thousand human inhabitants occupy a cluster of cement boxes off a potholed, meridianed boulevard called Broadway. During the war, barracked American soldiers convinced themselves that Tinian is shaped like Manhattan and so named the island's grid of roads after famous streets: Broadway, Canal, Forty-second. My own neighborhood, near the corner of Eighty-sixth Street and Riverside, translated on Tinian into an impenetrable dirt track. I was intrigued by the prospect of snakes in Tinian's Times Square—an intersection that turned out to be only slightly less weed-bound than Tinian's Upper West Side—but even if they were established there, the odds of my seeing them in all that foliage were remote.

"I've likened it to AIDS," Perry says. "It can be widespread in the population before it ever becomes apparent. Ebola virus is so apparent that it will always draw attention, but AIDS will cost more money and kill more people. The snake is like that. A little more money spent now would save much more later. California is spending twenty million dollars to stop Mediterranean fruit flies, and that's to protect avocados."

An epidemiologist studying malaria needs to know something about the mosquitoes that carry it. Likewise, Perry would like to learn more about the carriers of this herpetological contagion. What goes on inside a wheel well? What are the prevailing temperatures and humidities inside shipboard cargo containers? If a snake crawls inside, what are its chances of crawling out alive at the end of the trip? With that information, one could begin to rank commercial trade routes according to their likelihood of spreading brown tree snakes, and could likewise identify which islands, ports, and cities face the greatest risk of infection. "For example, we could find out what the risks are for the plane you came in on," Perry told me. "It was probably a DC-10. It's a seven-hour flight from Hawaii. If it flies at forty thousand feet, then the temperature is about minus forty degrees Celsius. That's pretty damn cold. The snake can't survive that. The question is, is it that cold inside the wheel well? Probably not—the landing gear has to be in working order. And we know that snakes get in there. I personally pulled one out of a plane bound for Japan four months ago."

Perry showed me his principal tool for this research project, a small rectangle of microelectronics called a data logger; it looked like a box of dental floss. The data logger is highly sensitive to its environment and capable of collecting nearly two thousand data points—temperature readings, humidity readings—over whatever span of time Perry designates. Afterward, the data can be downloaded into a laptop and analyzed. Perry has several data loggers, each with many thousands of miles logged. Some have spent weeks riding in the cargo containers of ships. Others have traveled in the luggage compartments of airplanes. Perry said, "We have a lot of information on luggage compartments. Snakes are very happy in there. They will come out alive." Perry has been impressed by the wide variation in climates. A cargo container might be baking hot inside if it is stowed on the sunny deck of a ship, or it might be pleasingly cool if it is stowed below in a dark hold. Wheel wells present further complexities. What cranny in this black nest of tubes and cables would a snake most likely be drawn to? Some planes have a starter motor that expels copious amounts of hot air through a tube in the wheel well. Others warm their passengers' feet with a series of hot-air tubes that pass under the cabin floor and may or may not extend into the wheel cavity. Every wheel well, every cargo container is an environment unto itself, a microhabitat as distinct as one branch of a tree from the next, or one species of disease-bearing mosquito from another. Recently Perry adapted his research into a paper—"Conditions facing airplane stowaways during flight: temperature in wheel-wells and cargo compartments"—for the journal *Aviation, Space, and Environmental Medicine.* His focus was not snakes, but humans: the handful of people each year who, desperate to immigrate somewhere, hide in the wheel wells of outbound jetliners. Virtually all of them die trying.

As vectors of contagion go, there is one glaring difference between mosquitoes—or rats or fleas—and airplanes: society as a whole approves of airplanes, whereas most people would happily see mosquitoes, rats, and fleas vanish entirely from Earth. This places the invasion biologist in an awkward position. Faced with an outbreak of a mosquito-borne illness such as West Nile virus, a traditional epidemiologist can openly propose—and political officials will readily consider—solutions that attack the vector directly, up to and including spraying insecticides in and over large swaths of a metropolis. An epidemiologist of an airplane-borne contagion of snakes has no such recourse. Natural science suggests one approach; social, economic, and political realities demand another. "We're very much on the interface of biology and policy," Perry said. "The perspectives are very, very different. Science is designed to produce the right answer. It's not designed to produce quick answers. And often what you're doing might need a quick answer. How do you make sure the snake doesn't get to Hawaii, Texas, Florida? That's an immediate issue: the snake can go out any day. And politics is very important. The main way the snake has gotten out is ships and planes. So commerce is involved. Money is involved. The problem is solvable, but if the solution is 'No planes anymore,' obviously that's not going to pass."

Science is a contest between two opposing but not entirely opposite phenomena. On the one hand is nature: the unruly physical world out there, filled with everything from quarks and ions to stars and galaxy superclusters; teeming (at least locally) with bacteria, plankton, grasses, birds, snakes, and, increasingly, people. It is a realm, in short, of objects and forces acting and acted upon in a manner that presumably would continue whether or not humans were around to contemplate them. Nature is what causes a tree to fall in the forest, even if no one is listening. Addressing this phenomenon, taking it all in, is the observant listener: the human ear, the human eye, the human hand—the roving, probing, attendant tools of a restless and inquisitive human mind. Over time, this mind has augmented its capacities by creating additional tools: sextants, telescopes, microscopes, particle accelerators, software programs, and a procedure loosely referred to as the scientific method, the proper rules of which have been debated ever since the human mind advanced to a point capable of considering such matters. Insofar as a single line can be said to separate the world of nature and the world of man, that line is exactly the thickness of the human skull. Outside, an infinite number and variety of dots; inside, a system that connects the dots. Outside: hippopotamuses, trees, deoxyribonucleic acids, evaporation, erosion, decay; inside, hypotheses, deductions, theories, predictions, calculations, conclusions. Outside, *silva rerum,* the forest of things; inside, dioramas, models, metaphors. Outside, true wilderness; inside, a map.

The science lies in learning, over and anew, how to tell the difference. Consider the case of the American astronomer Percival Lowell. In 1896, while peering through his twenty-four-inch-diameter telescope, Lowell observed what he firmly believed to be canals on the surface of the planet Venus. Lowell was already famous for spotting canals on Mars—proof, he convinced himself (and the lay public), of the existence of intelligent Martians, or at least Martians with shovels. His observations in 1896 led him to conclude that Venus was likewise inhabited. Where Lowell saw canals, however, fellow astronomers saw only a blurry white sphere; they could not replicate his observations. Neither could Lowell himself, until at last, in 1903, he saw the canals once more and even managed to draw a picture of them. Lowell died in 1916; the canals were never seen again. Only recently have scientists figured out what he saw. A team of ophthalmologists, examining Lowell's notes, determined that the astronomer had made the aperture of his telescope so small that he had unwittingly turned it into a mirror: what he thought were canals were, in fact, the reflected shadows of the blood vessels in his retina. He had stamped his eye on the cosmos and mistaken it for the real thing. It is every scientist's nightmare: to mistake inside for outside; to confuse the order one sees, or wishes to see, or even, as in Lowell's case, one actually *does* see, with what is actually there.

Invasion biologists readily acknowledge one such perceptual bias in their research. Much of what is known—and theorized—about ecologi-

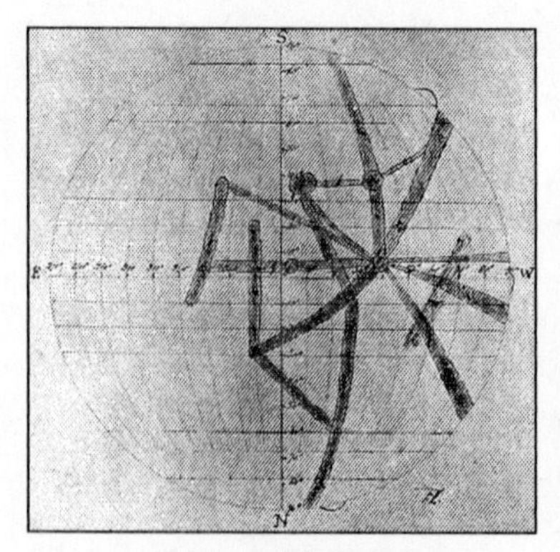

The "canals" of Venus

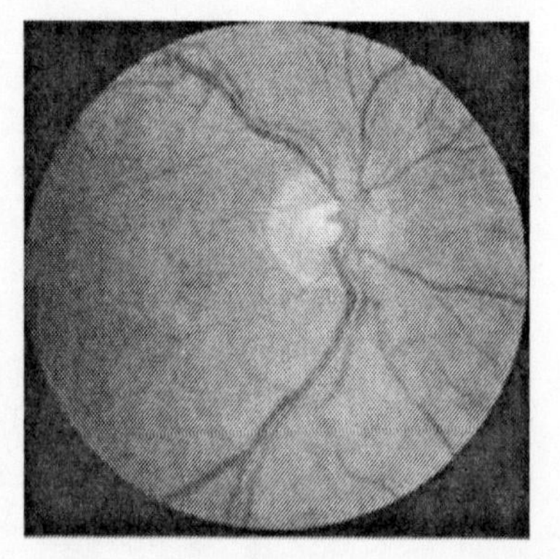

A human retina

cal invasions is based on case studies of successful introductions, because the failed invasions have disappeared and left scant record. However, it is hard to be fully confident about why things work when so little can be said about why they don't. The literature of invasion biology frequently makes reference to the "ten's rule," a highly anecdotal observation that one in ten invasion attempts results in successful establishment, and one in ten established invaders will become serious pests. Yet even this vague axiom is difficult to prove, as the bulk of scientific attention is given to the one exciting invasion, and little effort is given to studying, or even counting, the other ninety-nine attempts, or however many there may actually be. Daniel Simberloff, director of the Institute for Biological Invasions at the University of Tennessee in Knoxville, likes to emphasize the converse: most invasions *aren't* successful, and most established invaders *aren't* pests. Why not? Are the troublesome invaders biologically unique or similar in some measurable way, or are they simply statistical anomalies—the inevitable one in a hundred? "Without knowledge of the failures, it is impossible to answer questions at any of the levels listed above by observing numbers of successful invasions by species with particular suites of life history traits," writes Simberloff. That the most studied invasions tend to be the more noticeable ones, he adds, should give an invasion biologist pause before generalizing from any particular case study. The brown tree snake is iconic, but is it representative?

When Gordon Rodda surveys his field of research, he sees an epistemological minefield. Not only does he study the one-in-a-hundred successful pest of an invader (which may or may not be reflective of invaders elsewhere); he studies an invader whose surreptitious behavior makes it supremely difficult to study in the first place. Even seemingly straightforward acts of observation—for example, gaining an accurate count of the number of brown tree snakes in certain plots of forest on Ritidian Point, or of the numbers of geckos, skinks, and moles that inhabit those same areas and are available to the snake as food—are to Rodda's mind tripwired with potential viewer bias. "One thing I feel really strongly about is that ecologists generally don't understand how hard it is to get accurate measurements—and they understand better than the public does. The perception is, you just put a gauge on a pole, and that's it. But it's a physically and philosophically challenging problem."

Nor is it without consequence to the forests of Guam. One afternoon, Rodda invited the biologists of the Division of Aquatic and Wildlife Resources—the agency on Guam responsible for tracking and studying what few wildlife resources remain on the island—to the conference room at Ritidian Point for a viewing of some snake numbers he'd been crunching lately on his laptop. For several years, the DAWR biologists had discussed the possibility of reintroducing the Guam rail to a sixty-acre plot of native forest at the northern end of Guam. The plot, known as Area 50, sits on U.S. Air Force property amid the overgrown remains of runways. It is entirely surrounded by a Cyclone fence. I'd visited it one day with Rodda and Gad Perry—Perry driving a perilously decrepit car, Rodda crammed in the passenger seat, his cranium brushing the ceiling, wincing at the potholes. The forest itself, inside the fence, was a local anomaly of diversity: breadfruit and ironwood trees, and small cycads growing on a thin layer of soil overlying rough limestone karst. It was an ideal habitat for rails. The problem was snakes, which would have to be trapped out or eliminated, or their numbers somehow sufficiently reduced. Determining at what point the snake-substraction effort had succeeded and bird addition could safely begin was Rodda's project. Gathered in his conference room, the biologists plied him with questions.

"How do you know when the snakes have been effectively removed?"

"How many days do you have to go not capturing a snake until you feel confident you've captured most of them?"

"At what density of these snakes can the birds proliferate?"

"How low do we have to keep the snake population to allow the birds to recover? And how fast do you get to that point?"

In short, if the success of the snake, like that of many biological invaders, is partly a function of its silence—if there is an inescapable gap between the number seen and the number actually present—how does one begin to detect it?

Rodda is the scientist to approach with such questions. He has spent his years on Guam not only tracking and counting snakes—figuring out how many there are, or aren't, in a given area—but also evaluating and perfecting the best method of said counting.

"It appalls me the number of ecologists who use a technique because it's available, without first evaluating if it's the best one. If I see twice as many snakes today as yesterday, I would like to believe it's because there are twice as many snakes there." He added, "I tend to be more anal than most people about this stuff."

As it turns out, there is a best way to count snakes in the forest. Contrary to long-standing tradition, the proper method does not involve herpetologists in headlamps creeping among the trees at night and counting snakes in the boughs—one an hour, if you're lucky. This approach is flawed for the plain yet not-so-obvious reason (as Rodda discovered through a statistical comparison of his own nocturnal snake counts against those of Campbell, Clark, and Perry) that some people are better than others at seeing snakes. "The most important variable was not when or where you're looking, but who's doing the looking. This turned out to be a significant source of error." Instead, Rodda has begun to rely more on the use of snake traps to gain accurate counts. "Spaced twenty to thirty meters apart, you can catch up to a quarter of the snakes in an area in a single night, which is phenomenally good." The trap is a cylindrical cage roughly two feet long, resembling two heavy-duty wire-mesh wastebaskets attached at their top rims. A two-inch-diameter plastic cuff at one end serves as the door: the snake can enter, but a tiny wire flap, swinging inward, prevents its escape. The lure is a live white mouse, safely housed inside the cage in its own wire cabin with as much diced potato as it can eat in a week. Rodda designed the cage; he built and tested more than four dozen configurations before settling on this one. One early model had an 80 percent escape rate. Another set of tests examined the attractiveness of different baits. (Water, perfume, blood, catfish baits, and even commercially available snake lures hold nowhere near the appeal that a live mouse does; even traps into which a mouse has been inserted and then removed will continue to catch snakes.) To keep snakes from wandering too near any mode of outbound transportation, the traps are now a fixture along the Cyclone fences that surround Guam's airports, shipyards, and cargo-loading areas. The effort of checking them falls to the U.S. Department of Agriculture's department of Wildlife Services, which is designated to stem the tide of brown tree snakes from Guam. Once, the agency supervisor decided to tinker with the cage's wire entry flap, which in his opinion jammed too frequently. Rodda was unimpressed with the changes and, naturally, had a series of test-result graphs to back his claim. "Mess around with the flap, you do so at great risk."

The results of all the snake surveys, trap tests, gecko counts, and other data collected over the years on Guam have been fed by Rodda through various computer algorithms and transformed into a series of multicolored cross-referenced charts and graphs on his laptop. With his laptop plugged into an overhead projector in the conference room, Rodda gave the assembled biologists a picturesque tour of his information landscapes.

What it added up to was a sort of meta-ecology. How do you accurately measure what you can't see? How do you know what you don't know? How do you know when you know enough, or when you need to know more? The audience reaction that morning was quiet awe, as much for the boggling level of data analysis as for the sheer novelty of the effort.

Rodda's meta-questions apply not only to Guam, which already has the snakes, but to Micronesia, Hawaii, and every other location that has, might have, or thinks it one day may have brown tree snakes. In fact, these questions apply to all invasions and, Rodda emphasizes, to the even larger venture of studying biological diversity as a whole. Rareness and abundance, the twin measures of biodiversity and successful conservation, are matters of accounting. To calculate accurately how many, and how quickly, various species are going extinct around the world—owing to habitat loss, pollution, invasions, or whatever other reason—scientists must first determine the total number of existing species, roughly how many individuals of each species remain, and more or less where they can be found. Without proven and reliable accounting practices, all the counting in the world is fruitless.

"The public thinks the government is monitoring the populations of various things," Rodda told me. "It's just not true. By and large, nobody is measuring it. For most species in most places, nobody has a clue. For deer, elk, a few economically important species, yeah. But vast numbers of species, including important ones of economic value, nobody knows. And one of the reasons people don't know what's going on with population trends is because it's hard to make the measurements."

4

No organism, regardless of how small or brainless, aspires to be the last of its kind. Consequently, through evolutionary time, the species of the world have pursued all manner of strategy to expand their foothold on Earth. The seeds of the coconut can float for months on salty ocean currents, which explains their prevalence on tropical coasts and oceanic islands around the world. Snails and frogs from Cuba and Central America, borne aloft on the winds of tropical hurricanes, land in Florida with sufficient regularity that they have established nascent colonies in the Everglades. Ship captains have witnessed lush rafts of flotsam drifting far out at sea and harboring all manner of tagalongs: crabs, barnacles, field mice, boa constrictors, termites, geckos, squirrels, thirty-foot-tall palm trees. Charles Darwin noted that on several occasions the rigging of the *Beagle* became coated with the gossamer strands of spiders—thousands of spiders, the size of poppy seeds—blown several dozen miles offshore. "The little aeronaut as soon as it arrived on board was very active, running about, sometimes letting itself fall, and then reascending the same thread; sometimes employing itself in making a small and very irregular mesh in the corners between the ropes."

Near Cape Verde he examined a net of gauze attached to the *Beagle*'s masthead to see what other life might be carried in the winds. In addition to mosquito wings and beetle antennae, he recovered pollen grains from European trees; spores of fungi, lichens, mosses, and ferns; and the encysted forms of single-celled plants and animals that had been whirled aloft from dried-up ponds. Henry David Thoreau spent the last years of his life assiduously documenting the dispersion of seeds: the roving parachutes of milkweed and cottonwood that waft on summer breezes; the insistent burdocks and sticktights that will attach themselves to a dog's fur or a philosopher's pant leg and be carried for miles, all the way to his front door. "We find ourselves in a world that is already planted," he wrote, "but is also still being planted as at first." In recent years, whole subfields of science—aerobiology, radar entomology—have arisen to more closely study the wind-borne movements of plants and animals, using technologies more often associated with human traffic: airplanes, balloons, helicopters, Doppler radar.

Some organisms have linked their fate, or that of their offspring, to the movements of yet other organisms. Mussels of the genus *Lampsilis* spit the larvae of their young into the mouths of unsuspecting fish; the larvae attach themselves to the fishes' gills, where they draw nourishment until they are large enough to let go and burrow into the sand, miles from their parent. Young quahogs in New England disperse across the seafloor by clamping onto the legs of crabs that step on them—often with such bear-trap tenacity that they amputate the leg. Another tiny crab spends its entire life cycle traveling inside the respiratory tract of a lumbering sea cucumber; the pearlfish plays a similar role at the opposite end, darting in and out of the sea cucumber's anus generation after generation. Bees have mites; birds' nests have bedbugs. Evolutionarily, it pays to hitchhike. A migrating marsh bird is a more dependable mode of transportation than a random, wayward gust of wind. Burrs thrive in greatest profusion most closely to the footpath, where heavy traffic drops them—and will pick them up again.

And what carrier is more dependable—what organism travels farther and more frequently—than humankind? No species has expanded its range so willfully or with such success. We walked out of Africa north to Europe, east to Asia, and across ephemeral land bridges to North America and Australia. When feet proved too slow, we augmented: wheels, sails, rails, wings. Even today, when it sometimes seems that there is hardly anywhere new left to go, we keep moving, towing along everything necessary to create as good a life in our new homes as in the old ones. Standing outside Guam's international airport, on a bluff overlooking the island's industrial district, I began to grasp the scale of humankind's constant motion: square miles of commercial warehouses in which consumer goods—auto parts, floor fans, crates full of hand soap—await packing and shipment to smaller islands nearby and farther abroad. On any given day, Apra Harbor, the commercial seaport, is crowded with cargo containers the size of railroad freight cars. Break-bulk items too large for such containers wait outdoors on the pavement: stacks of lumber and long metal pipes; a truckload of acetylene tanks; fifteen new cars; ten historic powder cannons. The Fleet and Industrial Supply Center, the main cargo depot for the U.S. Navy at the northern end of the island, is a warehouse the size of a football field, filled with crates and boxes of all equipment vital to maintaining national security: ironing boards, clothes dryers, heavy-duty spoons, treadmills, earplugs.

The historian Alfred Crosby, in *Ecological Imperialism*, contends that agriculture and the domestication of livestock, the essential makings of modern humankind, were well in place by three thousand years ago. "The opposable thumb had enabled hominids to grasp and manipulate tools; in the Neolithic, these humans would reach out to grasp and manipulate whole divisions of the biota around them." We had begun to shape the wild into a space more familiar, to turn the unknown into the known. And the natural world—some of it—began to shape itself to us. Our plowed fields were an open invitation to opportunists: vetch, ryegrass, thistle, fast-growing plants already accustomed to colonizing lands swept clear by floods and fires. No sooner did humankind create the garden than she had to weed it. Animal pests—rats, mice, fleas, worms—were drawn into our domestic aura, as well as diseases unique to humans: smallpox, measles, dysentery, influenza. Consistently, nature has heeded the call of our campfires, our warm hearths, our alleyways and garbage heaps. Corn rot, bathroom mold, staphylococcus-resistant bacteria—we are their meal ticket to reproductive success. We are their movable feast. Now the brown tree snake too follows in the wake of civilization. Its migratory expansion shadows the human one; it has become what biologists refer to as a commensal organism—literally, a messmate—reliant on the habits of another species.

At this point a reader might reasonably wonder: How exactly does a brown tree snake find its way into an airplane wheel well? One morning, with that question in mind, I stopped by the Guam field office of the U.S. Department of Agriculture Wildlife Services. USDAWS consists of a dozen Jack Russell terriers, specially attuned to the scent of brown tree snakes, and a score of trained dog handlers. They have placed nearly two thousand snake traps in and around the island's seven ports, and they regularly check them. In addition, the dogs and their handlers visit the warehouses and cargo-loading areas as frequently as they can, in an effort to sniff out any snakes that have slipped past the gauntlet of traps and threaten total escape. Since its inception, in 1993, the unit has caught upward of fourteen thousand snakes, the overwhelming majority of them in traps designed by Gordon Rodda. The pop singer Mariah Carey, impressed by the plight of Guam's birds and the interdiction efforts of the dogs, has donated several Jack Russell terriers to the agency. She owns several herself, including one named Guam.

The USDAWS field office is located within the secured boundaries of Andersen Air Force Base, in a concrete warehouse on the outskirts of a small metropolis of military administration buildings. The location is ideal, as fully half the snakes caught trying to escape Guam over the years have been caught on Andersen property. The reason for this is plainly evident to a visitor standing on the tarmac: the base, its administration buildings, hangars, cargo heaps, airstrips, and the Cyclone fence around the perimeter are entirely surrounded by forest, hundreds of acres of prime brown tree snake habitat. A nocturnal tree snake runs on a basic cerebral program: move, climb, avoid light—three simple commands that, over evolutionary time, have reliably led the organism to food and shelter. So it isn't much of a trick for a snake to find itself in the shadowed midst of boxes and pallets, or ascending a treelike wheel strut, or amid the viny hydraulic lines inside a wheel well—and to then disappear. Once, in a short piece of pipe hardly three inches in diameter, Gordon Rodda found not one but eleven brown tree snakes in hiding, two of them well over four feet long. More than one snake has been found on USDAWS property: in one of the agency jeeps; in an inspector's boot; even, once, inside Danny Rodriguez's electric typewriter. Rodriguez, a former marine and F-18 mechanic, is the USDAWS field manager. At least one night a week he or another inspector hops into the company pickup to patrol the three miles of chain-link fence that encircle the military runways and separate them from the forest. Brown tree snakes will climb anything vaguely treelike—airplane wheel struts, but also Cyclone fences, which makes this fence perhaps the best place to try to intercept snakes before they reach the base itself. Rodda's snake traps are attached to the outside of the fence, about halfway up, one every thirty yards or so. When Rodri-guez spots a snake, either in a trap or entwined in the Cyclone fencing, he takes hold of it and pulls its head off.

The dogs serve as a last line of defense. They have caught snakes on the brink of export: in the power-steering unit of an outbound military truck; in a pallet of bombs destined for California; in a U.S. Navy SEAL trailer headed for Tinian. They have nabbed pregnant females—snakes containing five, six, eleven eggs. But the sun beats down, a dog's enthusiasm wilts, and the snake's instinct for seclusion is superb. Inspectors in a loading bay once watched a snake drop out of a military truck and disappear into the chassis of another, and still they could not find it. (It reappeared the following day, crawling around the legs of the driver.) All the while, the volume of outgoing cargo remains immense. The chief enemy of USDAWS is not biology, it's statistics. Scientists employ the term *propagule pressure*—the success of a seed owing to the sheer weight of numbers. One might, with Thoreau, picture the downy tufts of milkweed that in late summer spill from their pods onto the breeze. "The calm hollow, in which no wind blows, without effort receives and harbors it." One might notice, as I did on the day I visited, an enormous C-5 cargo plane parked at the end of the runway, where it had sat for days undergoing maintenance—its leggy wheel struts inviting and unattended, perhaps like the vectors that are thought to have borne several brown tree snakes to the air force base in Honolulu.

One night toward the end of my stay on Guam I looked out the window of my hotel room and beheld the most common manifestation of *Boiga irregularis:* total darkness.

Somewhere out there, the power had gone out. All the streetlamps were dark; the strip malls had shed their glare and receded from sight. The only visible lights belonged to pairs of headlights streaming up and down Guam's blackened main boulevard. The next morning, when the power was on again, I turned on the television news and learned that the outage had been caused by two brown tree snakes that had crawled onto power lines and cut electricity to the entire island. That this was newsworthy was itself noteworthy, as the phenomenon occurs almost daily on Guam: brown tree snakes induce more than a hundred outages a year as they cross power lines or slip into substations, perhaps in pursuit of the sparrows that nest there. Two snakes in one day was a new twist, though, and the incident presented me with a good opportunity to actually see one. I hurried over to the headquarters of the Guam Power Authority, where the press officer agreed to show me the culprits. She went to her office and returned a few moments later with a pizza box. She flipped it open to reveal two snakes, each roughly three feet long. The true color of their skin was now virtually impossible to discern, since it had turned mostly black from the high voltage and in some places was flayed off altogether.

My closest encounter with a live brown tree snake, however, came late one morning in full daylight, on an outing with inspectors from USDAWS. They were patrolling a densely forested area a few hundred yards from the international airport from which I would soon depart, and they were checking the numerous snake traps that had been placed there to capture would-be emigrants. The district supervisor led the way around trees and through thickets of chest-high sword grass. He sought out one snake trap after another, opened each one, and looked inside. On this day, at least, they were all empty. Finally, and perhaps keen to convey the effectiveness of his snake-control unit, he led me to a clearing beneath a large tree. A wire cage as big as a luggage trunk rested on the ground. Inside, in a writhing tangle, were at least a dozen brown tree snakes. Here was what I had been waiting to see: Hieronymus Bosch in a box. "We make sure we have some on hand for research," he said.

As the supervisor approached the box, a large snake lunged toward him, mouth agape, into the mesh wall of the cage. While the snake struggled to free its fangs from the wire, my guide lifted the lid of the cage and plucked the animal out with a pair of long metal tongs. He held it at arm's length and let it dangle for a moment—six feet of pure muscle coiling and uncoiling in midair. Its skin was neither brown nor blue, I noticed, but a murky green, like grass in Manhattan; the scales on its belly were yellow. "These ventral scales are kinda sharp," he said. "That's what allows them to climb." Then he handed the snake to me. At his instruction, I placed my thumb on a bony ridge at the back of its skull and, pressing down gently, clamped its jaw safely shut against my forefinger, as one might hold a garden hose. The rest of the snake immediately began to wreathe itself around my forearm and neck, squeezing with a gentle, almost clinical indifference, like the blood pressure gauge in a doctor's office. The tip of the snake's tail began to probe my right ear. I lifted its pebbly head for a closer look, as if I might catch a glimmer of intent. But its eyes were tiny yellow beads, slit vertically like a cat's and far more impenetrable.

At that moment I was of two minds. Although biologists do not like to talk about evolution in terms of "progress," I was struck most immediately by a sense of my own biological superiority. Any thoughts I was then having about the snake, it seemed clear to me, were a great deal more complex than any thoughts the snake might be having about me. I could both consider the snake and consider my considering of the snake, whereas the snake could only consider me as lunch. It seemed clear to me too—even if this calculus was less clear in the snake's unblinking eyes—that I was the organism in control: it was I holding the snake, not the other way around. At the same time, I was gripped by admiration. The snake was a marvelous work of biology—powerful, elegant, efficient. I was impressed by its opportunism, by its "strange craft": that an intelligence so minor could direct itself, unpremeditated, on such a profitable trajectory. As agents of homogenization go, it was not without appeal. I sympathized with it. I even felt a little sorry for it—guileless, cursed, outcast.

My sympathy, I knew, ran counter to customary thinking about ecological invasion. Conservationists and environmental scientists typically draw a firm line between two varieties of invasions: the unnatural kind, which happen as a direct or indirect result of human activity, and those that happen naturally without us. The latter category of invasion dates to the beginning of life on Earth; one recent paleontological study concluded that 95 percent of the animal species living today in South America evolved from species that traveled from North America—by wing, or by foot across a land bridge—millions of years ago. (The species that traveled in the opposite direction evidently fared less well: only 5 percent of North American species are evolved from southern ancestors.) Almost by definition, a species native to one place in the world—the Guam rail, or the now extinct bridled white-eye—is a species that evolved from an ancestor that somehow reached and colonized that isolated outpost long ago from somewhere else. That is natural—in contrast to the large and ever-growing number of ecological dislocations prompted, willfully or not, by our own motion.

The line between natural and unnatural, in other words, is us. On paper, the demarcation seems clear: If an invasion happened in prehistory, before the rise of human influence (whenever that was exactly), and left a mark in the fossil record, it is certifiably "natural." But in the first-person present—in this world awhirl with us, viewed without the aid of half a million years of hindsight—I was having a harder time of it. Some years back, the notion was floated that, with even the most far-flung wildernesses now tainted by the effects of human-induced global warming, nature, that separate and eternal realm untouched by our existence, had ended. But if nature was finished, what now was this thing that had wrapped itself so firmly around me, doing its Sisyphean best to finish me? Was it not nature? Was I not nature? That it could confuse a tree with a wheel strut, or a PVC pipe with a hole in the ground, seemed if anything to have worked in its favor. If the line we've drawn to distinguish natural from unnatural serves some human purpose, it is a line to which the snake—and every living inhabitant of the world, save ourselves—is entirely oblivious.

I understood that the snake had wrought tremendous environmental damage. And I understood that in an indirect way, as a consumer, traveler, and employer of the transportation the animal has latched onto, I bore some responsibility for its invasion. I could not condemn the snake without repenting myself. But for what exactly? Standing there, trying to draw reason from unforgiving eyes, I felt as if I had committed a most unoriginal sin.

Behind the building that Rodda and Perry call their office is a large yard of unmowed grass. Encircling both the yard and the building is a Cyclone fence, and encircling that fence is another fence. If years ago a spy had wanted to sneak into the building—into the steel-doored room where the submarine messages were intercepted, then into the sealed room where decoders feverishly decoded, and from there into the vault where the most precious secrets were locked away—this person would first have been required to navigate the two Cyclone fences and the vicious German shepherds that patrolled the alley between them. The dogs are long gone now, and in the yard inside the inner fence Rodda and Perry have added a fence of their own, a barrier designed not to keep things out, but to keep things in.

"There are barriers the snake can't get over," Rodda said, standing in the sunbaked yard. As important as population surveys and meta-analyses are in Rodda's larger scheme, it is the designing of fences that occupies much of his field time these days. Australia, he noted, has successfully fenced off thousands of square miles from introduced predators like feral cats and dingo dogs. The effort has been significant in protecting endangered kangaroos. "It's brute force. But it solves the problem exactly. If the problem is species moving into an area, you keep them out."

To halt brown tree snakes, Rodda said, there are two kinds of barriers one might erect. A long-term barrier, probably made of concrete, that could be set up around an airport or a nature preserve and could be maintained for several years. And a temporary barrier, something that could be deployed quickly around, say, a small cargo-loading area. The barrier might be an exclosure, designed to keep snakes out—useful for surrounding cargo being loaded into planes on Guam. Or it might be an enclosure, designed to keep any snakes that might be in the cargo from escaping on arrival into the wilds of Honolulu. In the coming weeks, American forces would be taking part in an annual multinational military exercise—held that year on the northeast coast of Australia, home not only to brown tree snakes but also to taipans, brown snakes, black tiger snakes, and several other candidates for the world's deadliest snake. Soon, mountains of cargo from Guam and throughout the Pacific would travel to Australia, sit for long stretches within crawling distance of the forest, and return: boxes, barrels, crates, cartons, pallets, trucks, and the associated dark, snake-friendly nooks. Among the cargo would be a portable barrier created by Rodda and Perry. "It's lightweight, inexpensive, and easy to transport and build," Rodda said. "It's designed to be used for a few weeks at a time. We tested the main idea. This is our final design."

He stopped in the middle of the yard, beside a roughly circular barrier some fifteen feet in diameter. The walls were waist-high and made of panels of heavy cloth, each panel attached to the next and pegged to the ground by a steel rod encased in PVC pipe. The cloth panels were not vertical but were angled inward; the whole contraption resembled a large canvas pup tent with its dome lopped off. The enclosure had been up for about a year, Rodda said. The panels were made of shade cloth, the sort used in greenhouses, because the material doesn't fall apart under the intense sun of the tropics. Also, the cloth surface was pocked with tiny holes: large enough to be porous to wind (of which Guam has plenty), but too small for a snake to squeeze through. "We've found that snakes can go through holes so small you'd think it's not possible. The snake keeps impressing me. It's a machine for getting places."

Gingerly, Rodda stepped inside the enclosure. A red fifty-five-gallon drum sat in the middle, on end. Somewhere in the ring there might be a snake creeping about, left over from a recent field experiment. Rodda wasn't sure. The snake is so cryptic that even if it is wearing a radio transmitter in a confined patch of lawn grass, a herpetologist—or, preferably, an assistant—must crawl around on all fours to find it. Rodda stopped to admire his handiwork. The design was deceptively simple, belying innumerable insights into brown tree snake behavior. In one early experiment, Rodda would dump thirty snakes inside the enclosure, then return the next day and count how many were still there. On other occasions, he or a colleague would sit on the barrel wearing night-vision goggles and watch what happened.

"As the night progresses, the snakes start trying to climb, and they're trying harder and harder. The only ones that can get out of here are two meters long or longer." Such specimens are rare, he said, maybe 1 or 2 percent of the population. Earlier experiments had revealed that the angle of the barrier wall was critical: the snakes tend to crawl up to the very edge of a surface and climb it; morever, they can free-stand up to two-thirds of their length. The backward angle causes them to collapse back under their own weight. In all, Rodda found, the snakes would spend only 3 percent of their time trying to climb the barrel; in contrast, fully half of their escape attempts were focused on the corners where the barrier panels came together, their best hope for a vertical ascent. "If you give them a ninety-degree corner, they can climb it. So it's very important that you have more obtuse angles. That's something we wouldn't know unless we spent hours sitting out here or watching videotapes from indoor experiments."

Back inside the building, between the conference room and the kitchen, is another room, sealed off from all natural light and tuned to an artificial cycle of day and night. Here, Rodda and Perry were testing designs for a more lasting barrier that, if effective, might be used to enclose several forested acres on Guam, from which brown tree snakes then could be trapped and removed and in which the remnants of Guam's bird population might be encouraged to recover. In the room, along two of the walls, Rodda and Perry had built a series of six wooden stalls, each with its own door, like walk-in phone booths. In each one, Rodda and Perry had constructed a different experimental design for a snake-proof barrier. Most of the test barriers were variations on a theme: a wall about chest-high with a wide, protruding ledge at the top, to keep the snake from simply slithering up and over, and an electrified wire or series of wires running along the top ledge to dissuade the most ambitious snakes. Brown tree snakes would be tossed into the stall to probe the weakness of the barriers. With the aid of infrared video cameras, one mounted at the top of each cubicle and aimed downward, Rodda and Perry could watch the confrontations and try to improve on each barrier design.

"We keep assuming snakes perceive the world the way we do," Perry said in an adjoining room, "but there's more and more indication that they don't. Put a brown tree snake in a darkened room: if there's any imperfection on the wall, they'll go for it. Maybe they can see infrared, I don't know."

Rodda said, "We're trying to get a feeling for how much it takes before what you want stops working. It's a balance between testing the most extreme conditions and making our lives as difficult as possible."

Perry popped one videocassette after another into a tape machine and watched as black-and-white video segments appeared on a television monitor—a sort of greatest hits of brown tree snake behavior. Long snakes, short snakes, snakes creeping to the wall and rearing up, snakes falling back in confusion from the electrical wire, snakes crawling over it, snakes in artificial daylight, snakes in total darkness. In one tape segment, the most advanced barrier was being challenged by a particularly large brown tree snake. "This one is what we call a honker," Perry said. "It's over two and a half meters long. In the population it's very rare. In our studies, they're all male. But even with the electricity off"—a common occurrence, owing to the snakes' knack for causing power outages—"only these large ones get out. Even if the worst happens and the barrier is not powered and the snake gets out, it may eat something, but it's male—it won't breed. We're very, very happy with this."

Perry's favorite tape is dated April 21, 1995. He slipped it into the video player and fast-forwarded to 9:08 p.m. A bird's-eye view appeared of cubicle No. 1, an early test model featuring a one-meter-tall barrier with a single electrified wire running along the top. On the floor, in total darkness, a honker was considering the possible avenues of escape. The snake crawled up to the barrier and rose onto its tail like a charmed cobra, swaying, leaning back, sizing up the barrier. Its actions might have appeared sinister had the tape not been playing at double speed; instead, the events had a Chaplinesque quality. At last, with its head alone, the snake found purchase on the ledge above the wire and in no time had pulled up the rest of its body, oblivious to the principles of electricity. The barrier had been surmounted, but escape was not yet complete. Earlier, when Perry had exhibited cubicle No. 1 with the lights on, he pointed out a piece of masking tape that had recently been laid to cover the hairline crevice that runs from floor to ceiling along the edge of the door. The need for this tape was now clear. Although the crevice was many times narrower than a snake, to this particular animal it must have yawned deliciously wide. Stretching over from its ledge, the snake pressed its body against this narrow crack and, using an impossibly slim door hinge for leverage, climbed the wall vertically, out of camera range. It was at this point that the video image began to shake violently. The snake, Perry later surmised, had begun climbing the electrical cable leading to the camera. Suddenly an enormous reptilian head loomed on-screen: out of focus, tongue flicking, one impassive eye considering another. The head disappeared; a long body trailed past the camera; the tip of the tail came and went. Time: 9:17 p.m.

"Well," Perry said, "we learned from that."

Paradise in Sight

5

What draws the traveler to an island that is farther from anywhere than anywhere? Do you aim to come, driven by a vision, with your destination clearly in mind? Or do you arrive by happenstance, like driftwood or a wafting seed?

When I was a boy, my favorite book was *Kon-Tiki,* the true-life adventure of Thor Heyerdahl as he sailed across the South Pacific on a balsa-wood raft. Mostly I just looked at the pictures, especially the one on the cover, a photograph of the raft seen from a distance: embattled by waves, with a tiny figure—a sunburned Heyerdahl—high in the rigging pondering the future of his waterlogged and slowly disintegrating craft. I imagined him alone, an island of man adrift on the blue expanse.

Heyerdahl, a Norwegian explorer and self-styled anthropologist, had spent a year in 1937 on the remote island of Fatu Hiva, in the Marquesas, deep in Polynesia. He had intended to study how animals had colonized such faraway and desperate terrain. Instead, he became fascinated with the people. Where were they from? How did the first settlers get there? Polynesian legend described a journey of intent, a steady, deliberate migration eastward across the Pacific that spanned thousands of years: first from New Guinea and the Solomon Islands east to Fiji, Tonga, the Cook Islands, and Tahiti; from there east to Easter Island and southwest to New Zealand; and at last to the Marquesas and fifteen hundred miles northeast to Hawaii. Heyerdahl was skeptical: canoe voyages on the open ocean, west to east against prevailing currents and trade winds, unguided by nautical instruments, landing with pinpoint accuracy again and again? Perhaps instead, Heyerdahl proposed, Polynesia was settled from the east, inadvertently, by South Americans blown adrift.

He conducted an experiment. He built a balsa-wood sailing raft, the *Kon-Tiki,* modeled after those he'd seen skirting the Pacific coast of

South America, and in 1947 he set off on the currents from Peru toward Polynesia. Heyerdahl was not alone, as I'd imagined at a young age, nor was he exactly aimless. On closer inspection the book's cover photograph reveals a second figure below Heyerdahl in the rigging; a third companion has rowed out in a rubber dinghy to snap the picture; and three more sit unseen in the raft's cabin, one reading Goethe and another charting their course with a sextant. In the end, after a hundred and one days and nearly missing their target entirely, the *Kon-Tiki* foundered on the reef of Raroia island. (Recently, a Frenchwoman sailed the same route alone in only eighty-nine days, on a twenty-five-foot Windsurfer.) Heyerdahl, and his worldwide readership of millions, were convinced: Polynesia was settled by accident, by human fluffs of dandelion.

This conclusion did not sit well among Polynesians, nor among Hawaiians to the north, who felt that their history and sailing prowess had been called into question. In 1973 a Honolulu anthropologist and two friends set out to prove Heyerdahl wrong. Their historical research found that early European visitors to the Polynesian islands in fact had witnessed enormous canoes—double hulls carved from tree trunks, with sails woven from coconut fronds—plying the open ocean. The newly formed Polynesian Voyaging Society decided to build one, following traditional designs and with modern materials. Then they would sail it from Hawaii to Tahiti—retracing, backward, the legendary path of Hawaii's settlement. For a navigator they recruited Mau Piailug, a forty-two-year-old native of the Caroline Islands in Micronesia. Piailug was one of the last masters of the dying art of open-ocean navigation: where Heyerdahl required a sextant, Piailug would refer to the drift of flotsam, the flight paths of birds, the patterns and reflections of clouds, the position of stars as they rose and set on the horizon. His crew set sail from Maui on May 1, 1976, and arrived without incident thirty-one days later on Mataiva, in the Tuamotus. They celebrated the Fourth of July with a traditional feast of poi and roast pig, then sailed home. Years later Piailug told a Honolulu newspaper, "When I go in the ocean, only one thing inside my mind: I like find the land. I no like miss."

The Polynesian Voyaging Society repeated the feat in 1980, this time under Nainoa Thompson, who had sailed with Piailug in 1976 and later trained with him. In 1985 Thompson undertook a two-year, sixteen-thousand-mile voyage through the island groups of Polynesia—again without instruments and, contrary to Heyerdahl's conviction, despite prevailing trade winds. Inspired by the example, other Polynesian islands built their own voyaging canoes. In 1995 three Hawaiian canoes sailed south to the Society Islands, met a contingent of three other South Pacific canoes, then sailed to the Marquesas and north again to the Big Island of Hawaii, for the first time retracing the original route of discovery of the Hawaiian Islands.

This historic voyage encountered only one significant problem. Three days before the canoes were due to arrive home, the crew radioed Honolulu with news that they were suffering the painful bite of some undetected insect. Entomologists were consulted. The suspects were narrowed to three: the nono, a tiny, vicious blackfly that plagues beaches of the Marquesas and doomed those islands' resorts; the punkie midge, a no-see-um half the size of the nono that also haunts the Marquesas; and a biting midge known in French Polynesia as the white beach nono. The first is native to the Marquesas; the latter two arrived in Polynesia sometime between 1920 and 1950. All three are functionally identical: with mouths like scissors, they bite holes in the victim's skin, producing welts that if scratched can quickly fester. A population of nono flies can inflict five thousand bites in an hour. After securing a blood meal, the fly retreats to a crevice in any nearby decaying organic matter—waterlogged wood, a coconut husk, the hull of a Polynesian voyaging canoe—to reproduce. Its larvae emerge several days later to begin the cycle anew. With these facts in mind, Honolulu newspapers treated readers to several days of midge coverage, disagreeing only as to whether the midges posed a worse threat to the tourist industry than the brown tree snake, or merely an equivalent one. Cynics whispered of a midge conspiracy, of midges planted aboard the canoes by an environmental group eager to publicize the dangers of alien species in Hawaii.

Riding a strong headwind, a biting midge can cross fifty miles of open ocean. To forestall such an event, and despite prime sailing conditions, the six voyaging canoes were ordered to a halt some two hundred miles south-southwest of Hawaii's Big Island, Hawai'i. The following morning, after much wrangling over which government agency should take responsibility for a nuisance insect that is neither an agricultural pest nor, strictly speaking, a public-health threat, and which in any event now sat in international waters, a Coast Guard plane dropped thirty-six aerosol cans of pyrethrin, an insecticide made from daisies, into what were now heavy seas. The canoes' holds were emptied, clothes and equipment sprayed with disinfectant, the hulls scrubbed four times with seawater, the sails keelhauled. Everything organic was tossed overboard: religious carvings, palm-frond baskets, breadfruit seedlings wrapped in coconut husks, and traditional foods like sweet potato, taro leaf patties, and poi—which crew members perhaps were happy to see go, as they later confessed a preference for sausage and Spam. Inspectors from the Hawaii Department of Health then boarded the canoes and sprayed everything again. A series of triumphal celebrations had been planned to mark the voyage's end. Instead, the canoes were escorted into Hilo Harbor, sprayed once more, and enclosed in fumigation tents. A crowd of two dozen, including several customs and immigrations officials, greeted them.

Health and agriculture inspectors still do not know what sort of midge they nearly encountered; the cleaning, spraying, scrubbing, and fumigation had left no trace of them. Other survivors were discovered, however, including four species of fly, two species of ant, a cockroach, two spiders, a book louse, a parasitic wasp, a beetle, several snails, some live shrimps, a gecko, two species of eye gnat, and a scale insect that in some parts of the world is considered a serious agricultural pest. Some time afterward the chief of the Hawaii Department of Agriculture was heard expressing nostalgia for the days when state inspectors could walk down the aisles of arriving aircraft and fumigate freely, as they still do in New Zealand. As for the canoes, they reached their destinations. The Hawaiian crews sailed to warm homecomings on Maui, Molokai, and Oahu. The Cook Islanders disembarked, showered, and dried off with one hundred and fifty donated towels. The Tahitian voyagers, though they had a fine canoe, never arrived; in fact, they never embarked. They had failed to apply for U.S. visas, and were gently advised to stay home.

The Hawaiian archipelago has long been thought of as a sort of laboratory of the pristine. It is an ecological jewel: 90 percent of Hawaii's native species—the species that were in Hawaii before the arrival of humankind some fifteen hundred years ago—are in fact endemic, found nowhere else on Earth. As such, almost since Captain James Cook stumbled upon them in 1778, the Hawaiian Islands have attracted scientists keen on understanding how plants, animals, and ecological communities arise and sustain themselves in such splendid isolation. And not only scientists are attracted. On average, and in addition to the human travelers and curiosity seekers, five new plants and twenty species of insect, plus the occasional pathogen and predator, are established in Hawaii each year. According to the Nature Conservancy, a "silent invasion of pest species" is under way.

Some are agricultural pests, such as the sugarcane aphid, accidentally imported a few years ago in a shipment of golf-course sod. Others, like the poisonous Amazonian tree frog, which now inhabits the upland forests of Oahu, pose a threat to humans and animals alike. All told, counting the nine hundred nonnative plants now growing in Hawaii, the foreign species approach in number the indigenous ones. The brown tree snake is only one of the more spectacular of the state's myriad concerns. "It is quite an exchange and bazaar for species, a scrambling together of forms from the continents and islands of the world," Charles Elton wrote of Hawaii in 1958 in *The Ecology of Invasions*. "It's an ecological disaster area," the biologist E. O. Wilson had told me over the phone before I left New York. "It's the invasion capital of the world." Who wouldn't be curious to see that?

In recent years, the effort to stem the tide of alien species into Hawaii has come to involve a tangled web of agencies and organizations: the U.S. Department of Fish and Wildlife; Hawaii's Division of Forestry and Wildlife; the Hawaii chapter of the Nature Conservancy; the Armed Forces Pest Management Board; the Killer Bee Committee. Much of the frontline legwork, however, falls to the several dozen inspectors—human and canine—employed by the Hawaii Department of Agriculture's Animal and Plant Health Inspection Service. The bulk of their time is spent standing in the river of passengers, cargo, post office parcels, and express-mail packages that flows steadily into the state, keeping eyes and noses open for unapproved nonnative organisms. Theirs is a ceaseless task of small triumphs, which the agency dutifully and proudly records at its headquarters, a small building that I eventually found tucked behind a warehouse in downtown Honolulu. The tally for one recent year added up to a hundred sixty-one animals: forty-eight toads and frogs, ten turtles, one mynah, six Jackson's chameleons, seven iguanas, seventy-five other lizards, four snakes, two hamsters, three ferrets, two hedgehogs, one hermit crab, one squirrel, and one possum. Most animals seized by the agency are eventually shipped back to the mainland and distributed to zoos. Any creatures required as evidence in court cases—an illegal white tarantula that someone had left in a paper bag at the airport; a pair of piranhas, seized from the home of a pet-store manager, that had recently finished spawning in their holding tank—were maintained behind a door at the back of the office, in a dank purgatory of glass terraria and clacking, burbling aquariums.

When one of Hawaii's various ecological invaders exceeds its stay in the animal room, and if no zoo expresses interest in its fate, the creature might well end up in one of the large glass display cases set up in the visitors' lobby to exhibit the state's plight. Here, stuffed and mounted behind glass, were ferrets, boas, bats. Here too were seven of the brown tree snakes that are known with certainty to have entered the state since 1981. For my viewing benefit, an agent opened one of the glass doors and withdrew from the cabinet a large jar of formaldehyde, which he set on a nearby table. He lined up five more jars alongside this one. White labels on each jar noted the length of the specimen within and its date of capture. The fifth jar, dated September 3, 1991, held the remains of two brown tree snakes found that day: one, dead less than twenty-four hours, on the runway of Honolulu International Airport; the other, stunned but alive, under the wing of a military transport plane that had arrived the previous evening from Guam. Except for one or two squashed heads, all the specimens were intact, each curled in its container of fluid, the lot of them suspended in a row like conspirators on a public gallows. They were slim and small, none longer than a meter; over time, their color had faded to a sort of pickled white. Even inanimate they retained an air of permanent menace, their mouths agape, their eyes clouded and unfathomable.

Traditionally, natural scientists sought to understand nature by seeking out "natural" enclaves of it, areas far from the footsteps of humankind, well off civilization's trodden path—in situ isolates of Eden. Beginning around the 1920s, however, and with increasing confidence through the century, naturalists embraced in addition a more interventionist approach. No longer content to simply study a lion at arm's length, they would also "twist the lion's tail," as Francis Bacon put it. They would experiment: in the field, in the laboratory, and everywhere in between. "Tinkering, observing nature out of sheer curiosity, and armchair theorizing as the early nineteenth-century philosophical naturalists did, have fallen out of favor in ecology," a commentator notes in a recent issue of *The Quarterly Review of Biology*. For Charles Elton and subsequent biologists, the innumerable examples of biological invasion around the world provided a ready supply of field studies, each one a potential test case of ecological theory. "The weeds and cultigens, scorned as 'unnatural,' yet may tell us a great deal about nature," wrote the natural historian Marston Bates in 1956. "Their behavior is part of an unplanned but nonetheless significant series of gigantic experiments which, by the very alteration of the geological sequence of events, may teach us much about the operation of that sequence." Even in that era it was apparent that Hawaii represented a laboratory full of such experiments. It was a microcosm of microcosms—"probably the most changed of the major faunas through man-dependent dispersal," Bates noted. The same holds true today, more than ever. For better and worse, Hawaii is two test tubes at once: in one, a study of the evolutionary effects of biogeographic isolation; in the other, a runaway study of the consequences of breaching that isolation. "It could be that Hawaii is a lost cause," a national-park manager in Hawaii confided to me one afternoon. "We'll see. In the meantime, it's a great experiment."

Going there, I suppose, was a sort of experiment for me as well. If, as scientists say, the danger of biological invasion is the homogenizing veil of sameness it imposes on the landscape, I had met one of its weavers: in the forests of Guam; and here again in Honolulu, through a glass jar, I looked it in the eye, dim and thoughtless though that eye was. What I desired now was its obverse: nature uncloaked—splendid, wondrous, naked. I wanted to see what's at stake.

Well, that is only partly true. I had hardly elbowed my way through airport baggage claim when I became dimly aware of another impulse for visiting Hawaii, this one more reptilian than noble, and not terribly novel: I just wanted to see the place. Soon after my arrival, while poking through the library of the natural history museum in Honolulu, I had come across a travel journal from the 1920s. The authors, a pair of British medical entomologists, had journeyed through Polynesia and Melanesia surveying the local population of bloodsucking insects. They soon discovered that many of these insects in fact were relative newcomers to the islands, carried there inadvertently by the investigators and adventurers that had preceded them. The louse is "a recent introduction into the New Hebrides, where it is known as 'One big fella flea, 'e stink.'" The residents of another atoll had first made the acquaintance of the common flea only a hundred years earlier, in the 1820s—an event thereafter conflated with the arrival of Western explorers. "The placid natives of Aitutaki," the journal notes, "observing that the little creatures were restless and inquisitive, and at times even irritating, drew the reasonable inference that they were the souls of deceased white men." Now, standing in the lobby of the Hawaii Department of Agriculture and peering into the hollow eyes of that snake, I had the disturbing impression that I had seen a ghost.

6

If it is true that nature ends where humans begin, then nature in Hawaii first ended fifteen hundred years ago at Ka Lae, the southern tip of Hawai'i, the Big Island, at the foot of towering Mauna Loa. Ka Lae today is the southernmost point in the fifty United States, and it is a desolate place, a grassy, windswept knob of land that ends abruptly in a line of black lava cliffs above a roiling turquoise sea. No monument marks the precise landing point of the first Polynesian settlers, but over the centuries they would make their presence felt. They spread throughout the archipelago, using stone adzes to transform the lowland forests into fertile farmland, building domestic lives around taro, pig, breadfruit, and other plants and animals they had brought with them. Villages sprang up, of which, for a time, Ka Lae was the most populous. Kamehameha, who in 1810 became the first king of all the Hawaiian islands, is said to have fished here; Kamehameha II surfed nearby. Much later, after syphilis, influenza, and smallpox ravaged the population, and eruptions and mud slides devoured the town, nearly a century after Hawaii was forcibly annexed by the United States, some rich developers briefly considered building a spaceport at Ka Lae, a Pacific rival to Cape Canaveral. But all I saw on the day I visited was a herd of cows roaming beneath tall wind-power generators and a group of local fishermen sitting along the cliff on picnic coolers, casting into the waves.

From Ka Lae a bumpy dirt road leads north to Mamalahoa Highway, the two-lane road that encircles the island. Here a modern traveler faces the first of many choices. In Hawaii, directions are given not so much according to the points of the compass as by one's position relative to the trade winds. Are you heading to the leeward part of the island or to windward? Are you going to *kona* side or *hilo* side? Are you going to the beach, or do you want to get drenched? Ka Lae, dry in the rain shadow of Mauna Loa, is *kona* side. Turn left on the highway, roughly northwest, and one eventually reaches Kailua-Kona, a warm, cloudless seaside town notable for its rich coffee fields and, increasingly, golf courses, shopping centers, and sprawling resorts. I turned right, to *hilo* side: up the southern flank of Mauna Loa to the four-thousand-foot elevation mark—a third of the way to the summit—then down the other side toward the island's capital, named Hilo. The transition from *kona* to *hilo* occurs precisely at the crest of this road. One moment a driver is ascending through arid blue brightness; suddenly a wall of cloud looms ahead, and then you are in it, wipers swishing, driving downhill between walls of moss-draped tree ferns. I crossed this line dozens of times during my time in Hawaii yet never failed to be astonished at how a day could congeal into mist so instantly or dissolve again if I simply chose to reverse course.

Just south of the highway, straddling the *kona–hilo* divide, is Kilauea Caldera. The Big Island is a crucible of evolution; Kilauea is the flame of it, a straight-walled basin three miles across and five hundred feet deep. Inside, at one end, is a smaller crater, Halema'uma'u, the legendary home of Pele, Hawaiian goddess of fire. From such craters, on occasion, sometimes for months or years on end, lava is ejected hundreds of feet into the air—in molten sheets or in flaming ropes and ribbons—then falls spattering into the surrounding basin and cools into a rocky black crust. This happens so regularly that the floor of the caldera resembles a cracked and rumpled plain of black meringue; sulfurous clouds pipe up here and there, rising out of the basin to form an erosive volcanic fog, called, technically, vog. Mark Twain visited Kilauea in 1866 and, only hours after an eruption, spent a pitch-dark night stumbling across the newly crusted floor, searching for the trail. "I have seen Vesuvius since," he later wrote, "but it was a mere toy, a child's volcano, a soup kettle, compared to this." On occasion, lava spews from the summit of Mauna Loa, ten thousand feet higher up, but such events are hard to witness because the strip road that leads there has been buried by successive eruptions. Mauna Loa and Kilauea, as well as the satellite craters that surround them, branch from a single trunk of magma rooted in the seabed miles below. Lava wells up from the Earth's crust and out onto the Big Island, adding an inch of height to it each year. And each year, with the slow migration of the continental plates, the island drifts a little farther west off its stem. Already, from an underwater vent a mile to the east, a new island, Loihi, is taking shape, although scientists do not expect it to surface for several thousand years at least.

Psychologists have found that what people think of as the present actually encompasses about three seconds, roughly the span between blinks of an eye: time enough to think, *This is the present,* but not enough to think, *I'm still thinking about the present.* By that time, you are already into a new present, the old one having quietly subsided, a mote drifting down into the sediment of memory. Volcanic islands are like that: they rise up; they linger for a geological blink; they disappear beneath the waves. They go extinct. Of course, geological blinks are very long. A great many biological moments can be packed into a single geological one. In the seven hundred thousand years that the Big Island has existed, thousands of species have arisen on it, some of which went extinct even before scientists evolved to describe them, each species the product of hundreds or thousands of generations, each generation composed of countless individual lives, countless blips of empirical sensation—a magnanimous observer would say consciousness—each too brief and preoccupied to grasp the larger geological moment that contains it. And that moment is still young: the Big Island has perhaps thirty million years to go before it winds up like Ocean Island, the oldest and westernmost of the Northwest Hawaiian Islands, today little more than an atoll peeking above the waves like a periscope.

The occupants of the brief geological moment called the Big Island are concentrated in Hawai'i Volcanoes National Park, which unfolds for more than two hundred thousand acres around Kilauea Caldera. Opened in 1916, HAVO stretches from the summit of Mauna Loa down to Kilauea and fans out below that to the coastline, encompassing rain forests, ancient xeric woodlands, and a vast expanse of barren, sunbaked lava flows known as the Ka'u Desert. Within these borders reside one of the most extensive remnants of the archipelago's native flora and fauna. Park surveys list nearly five hundred native plants, including ninety-six species of fern, and eighteen native animals, eleven of which are officially endangered: four forest birds, a hawk, an owl, a goose, two sea turtles, a seal, and a hoary bat. One mandate of the park's managers is "to protect natural habitats from alien biota"—a suite of offenders that includes feral goats and pigs, mongooses, cats, rats, and hundreds of alien plants, fifty-six of which have been deemed especially pernicious. Although some species were introduced with the arrival of the first Polynesian settlers several centuries ago, most ecologists mark 1778, the year of Captain Cook's arrival, as the beginning of an accelerating decline.

"The early Hawaiians altered the lowlands considerably," Linda Pratt, a veteran Park Service botanist, said one morning during a stroll through one of the high-elevation forests. "But they had virtually no impact on these upland systems. What we're looking at today is two hundred years of change, very recent and passing away before our eyes."

What we were looking at was a stretch of forest on the dry side of the *kona–hilo* boundary: a sunny, open woodland and one of Pratt's favorite sections of the park. Pratt has worked at the national park for more than two decades, monitoring the advance of alien plants and compiling a detailed history of the introductions; her husband is a park ornithologist who has studied the Big Island's birds for at least as long. On this morning she had tucked her hair—long and brown, with faint wisps of gray—under a green Park Service cap. The park entomologist, David Foote, accompanied her. The two had spent the previous hour on the other side of the *kona–hilo* boundary, in a wet forest, and they still wore the blue rain slickers and black, knee-high rubber boots necessary for such an outing. The gear would be thoroughly scrubbed upon return to park headquarters. "Anybody who works in conservation in Hawaii is aware that they personally could be the one who carries some nasty alien plant into a native forest," Pratt said. A few years earlier, near the finish line of the Kilauea Crater Run, an annual footrace around the perimeter of the crater, park biologists had set up a booth where runners could have their sneakers scrubbed clean of alien seeds. The seeds were saved and later grown in the park greenhouse. Sixteen different plants sprouted, twelve of them common, and one, an African grass, an alien weed.

"Things will always be arriving," Pratt said. "The key element here is the time frame and the number. If all the people disappeared from the planet tomorrow, there still would be a very low rate of plant and animal introduction into Hawaii, say one every twenty thousand years. But the current rate is *hundreds* per year. To me, that makes a huge difference. Of the hundreds of plants that arrive in Hawaii, maybe only one, or none, could have gotten here on their own."

The difference between a *kona*-side forest and a *hilo*-side forest amounts to about a hundred inches of rainfall a year. That is to say, the Hawaiian rain forest, in true rain forest fashion, is very wet. Tree ferns spout from the ground like the tops of giant pineapples, an impenetrable field of them, their wide fronds overlapping and laden with glistening mosses. Fallen logs give rise to leafy lobeliads, trunks are entwined in flowering lianas, epiphytic liverworts sprout from the junctures of boughs;

barks bear slime molds, mosses have mildews. Everything is a staging ground for something else: move over, grow faster, waste no space. All of it is drenched in mist or rain, the air thick with efflorescence and decay. What does not sprout in excess is rotting underfoot, degenerating into a black, pungent muck that threatens with every step to suck the boot from one's foot. In a biological sense, the Hawaiian rain forest is perhaps the safest in the world: no poisonous snakes to worry about, no lurking leopards or gorillas, no nettles, no mosquitoes—no native ones, anyway, and none in abundance above the four-thousand-foot elevation mark. Still, an ecologist's job is hazardous. "You get bruised, constantly slipping and falling," Foote said. "I've toyed with wearing safety glasses in here." One park technician nearly lost an eye when she fell onto a branch that cracked the lens of her eyeglasses. Another was confronted by her physician, who demanded assurance that the bruises on her body were work-related and not the result of domestic abuse. Foote said, "It's pretty miserable work. I try my best to warn applicants. When they call and ask what it's like to work here, I ask them what their interests are. If they say they're looking forward to getting to a beach in Hawaii, I almost won't hire them on the spot—we're so far from any beaches, and it's just miserable if you're stuck up in the rain forest and it's cloudy and rainy all day and what you came to Hawaii for was fun in the sun."

That was one of the principal virtues of the dry, *kona*-side forest in which Foote and Pratt now stood: warm sunlight. Stout trunks rose dozens of feet straight up. Pratt pointed out a stately koa, once the predominant hardwood tree in Hawaii but so scarce now after decades of logging that when the Polynesian Voyaging Society tried to build an ocean-voyaging canoe entirely of traditional materials, they could not find two koas big enough to serve as hulls. (They settled for two spruce logs donated by Alaska.) Pratt pointed out a flaky-barked soapberry tree and a red-blossomed o'hia, typically the first tree species to colonize fresh lava floes and ubiquitous in both wet and dry Hawaiian forests. The vista was refreshing, a vibrant antidote to the monochrome forests of Guam. But there was something else too: in the weeks I'd spent on Guam, in a landscape the brown tree snake had depleted of birds, I had forgotten that forests sing. Silence has a way of doing that, of blanketing not only sound but the very memory of it, until the quiet—static and empty—comes to seem like the natural order of things. The Hawaiian forest, in contrast, swelled with an aural dew of warbles, trills, and whistles. It struck me as the paragon of wilderness, the iconic setting one conjures when imagining nature in its purest form. It is a vision absent of snakes.

When conservationists envision the damage the brown tree snake might wreak upon Hawaii, their thoughts turn first and foremost to the islands' endemic birds, the most spectacular among them the family *Drepanididae*, the honeycreepers.

Had Darwin visited Hawaii, he would have loved the honeycreepers. Instead he went to the Galápagos, and the finches he found there—thirteen species, known today as Darwin's finches—became the emblems of evolutionary biology. Related to one another by a single ancestor blown adrift, the finches nonetheless vary widely in form, many of them involved in profoundly unfinchlike activities. One drinks the nectar of cacti; another draws blood from the backs of boobies; a third trims twigs with which it probes tree bark for grubs. They are aided in these tasks by beaks that, in their diversity, have been compared to a toolbox of pliers: straight needle-nose, curved needle-nose, parrot-head, diagonal, lineman's. Collectively, Darwin's finches have come to illustrate the arc of evolution by natural selection: how environmental forces—flood or drought, seasons of plenty or privation—can work upon subtle, often imperceptible genetic variations to bring about wholesale changes in morphology, ultimately giving shape to new species, each one better adapted than its predecessor to the current environmental conditions, however transitory those conditions prove to be.

As illustrative as are Darwin's finches, the Hawaiian honeycreepers are more so. For one thing, they are more numerous: forty-seven known species in total, all of them, like their Galápagos cousins, descended from a single wayward pair of birds. In recent years, experts in genetics have estimated that the ancestral pair, the first honeycreeper colonists, reached the archipelago between fifteen million and twenty million years ago. The time frame is noteworthy, because Kauai, the oldest island in the principal Hawaiian chain, is only five and a half million years old. The Northwest Hawaiian Islands—the Hawaiian islands of yesteryear, which long ago sprang one by one from the hot spot beneath Kilauea and today are faded and sinking embers far to the northwest—range in age from seven million to twenty-seven million years. That is to say, the honeycreeper family tree took root and begun evolving on islands that no longer exist. The first pair blew in, colonized, and, after generations, radiated into new species. From one of these, another pair flew or was blown to the next island, colonized it, branched out into new species—over and over, again and again, the unfortunate ones stranded on an island doomed eventually to subside, the lucky few hopping to the next, emerging island, until Kauai and Oahu and the rest of the archipelago that they now inhabit rose above the waves. The honeycreepers are Darwin's finches cubed.

And they look it. Unlike the Galápagos finches, which occupy the narrow wavelength between black and brown, the Hawaiian honeycreepers display the full spectrum of raiment: the vermilion i'iwi; the leaf-green o'u; the crimson apapane; the mamo, glossy black, with shoulders, rump, and thighs drenched in yellow. Each species often has close cousins on other islands, each admitting further refinement. The scarlet Hawai'i akepa, for example, differs only slightly in hue from the rufous orange Oahu akepa (now extinct), the dark orange Maui akepa (possibly extinct), and the pale olive Kauai akepa, while the various species of amakihi span the subtleties of yellow-green. But what sets the honeycreepers apart is their astounding array of bills. The beaks of the o'u, the koa finch, and the palila are parrotlike variations on short and stout; these beaks are employed, respectively but not exclusively, in capturing and crushing beetles, tearing open the seed pods of koa and mamane trees, and extracting the maggot-size germ from the seed of the naio tree. The bill of the akepa is short and mildly twisted at its tip, an adaptation that mystified ornithologists until the bird was seen prying open leaf buds to get at the insects hiding within. Perhaps most striking are the bills of the nukupu'u and the akialoa, two birds rarely sighted these days. Their bills, which are slender, exceedingly long, and delicately curved, prove supremely useful for withdrawing nectar, in the manner of hummingbirds, from the slender-throated blossoms of certain native lobeliads. Indeed, as there are no native honeybees in Hawaii, the birds are thought to be one of the primary pollinators of these rare flowers.

Collectively, the honeycreepers take their name in part from their habit of creeping around the trunks and branches of trees, prying into bark and under lichens for whatever caterpillar, grub, cricket, cockroach, millipede, or spider's egg might be found there. Yet for all the specialization their beaks afford them, the honeycreepers share a taste for nectar. The o'u is partial to the female flower of the ieie vine. The amakihi, its bill too short to drink from the slender-throated lobeliads, has taken to puncturing the blossom at its base and retrieving the nectar more directly. The akohekohe sometimes can be seen with its crest dusted with pollen after plunging headfirst into an o'hia blossom. Accordingly, the one anatomical trait the honeycreepers have in common is a tubular tongue: longer or shorter, wide or slender, the tip sometimes ending in a wick of bristles, like a paintbrush, but always the edges of the tongue curling upward, often overlapping, forming what looks like a canal or a canoe or a finely rolled cigarette. It is by means of this tongue that the honeycreeper draws nectar from the flower. Even those few species that have forsaken this ambrosia altogether, that feed exclusively on seeds or caterpillars, retain its memory in the furrow of their loosely curled tongues. It was the music of these tongues that spilled now from the canopy, showering like one of those Amazonian cascades that has as its source a dozen streams, each born in some prehistoric pocket of wilderness, each tumbling down the ages to join the others in a river of gathering force and swiftness that soon explodes from a great height upon the present moment.

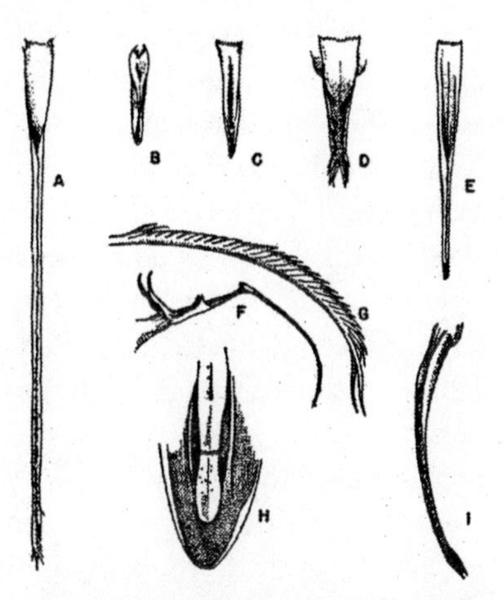

FIG. 16. Tongues of some species of Drepaniidae. A. *Hemignathus procerus*. B. *Psittirostra cantans*. C. *Loxops maculata bairdi*. D. *Pseudonestor xanthophrys*. E. *Himatione sanguinea*. F. *Drepanis pacifica*. G. Tip of same, enlarged. H. *Psittirostra bailleui* (tongue in place in lower mandible). I. *Ciridops anna*. A and B after Gardner; C, E, and H after Gadow; D, F, G, and I from Rothschild, drawn by Frohawk.

Or so it sounded to my expectant ear. Pratt smiled ruefully. The honeycreepers, she said, and the vast majority of the island's native songbirds, resided in forests elsewhere in the park, far out of earshot from where we now stood. She picked out a strain of song from somewhere off to the left. "That's either a Japanese white-eye or a European house finch," she said. She jabbed a thumb in the opposite direction. "And that's a cardinal over there. That's what gives New England forests their distinctive summer sound." So eager to catch a glimpse of unfamiliar nature, I had succeeded only in rediscovering my own backyard.

It is a unique trait of human beings that, however far we travel, we rarely go unencumbered. No other species packs a suitcase, or a lunch. That was the breakthrough of the Neolithic Revolution: the discovery of a reliable set of animals and plants—cattle, goats, pigs, horses; rye, wheat, barley, hops—around which to build an agrarian lifestyle. Those crops and beasts have since become regular members of the ever-expanding human road show. Maize joined the international troupe in the sixteenth century when European explorers discovered it in the Americas, where it had been domesticated for several millennia already; by the late eighteenth century it was traveling with English colonists to Australia. The historian Alfred Crosby notes that virtually every edible plant and animal described positively in *Letters from an American Farmer,* written in 1782 by the French settler J. Hector St. John de Crèvecoeur, in fact was of European origin. The pattern is even more entrenched today. According to a 1998 report from the U.S. Census Bureau, 98 percent of the American food system, worth an estimated eight hundred billion dollars annually, is derived from introduced species, from corn and wheat to cattle and other livestock.

No sooner do people export themselves than they begin importing other things: livestock, feed grass, lawn grass, ornamental shrubs, pet iguanas, pet canaries, cats. "I saw cats," Mark Twain wrote of Honolulu in 1866. "Tomcats, Mary Ann cats, long-tailed cats, bob-tailed cats, blind cats, one-eyed cats, wall-eyed cats, cross-eyed cats, gray cats, black cats, white cats, yellow cats, striped cats, spotted cats, tame cats, wild cats, singed cats, individual cats, groups of cats, platoons of cats, companies of cats, regiments of cats, armies of cats, multitudes of cats, millions of cats, and all of them sleek, fat, lazy and sound asleep." Five thousand years ago, after Eurasians had settled the Australian continent for tens of thousands of years already, they brought over their dogs. Australia today is plagued by feral camels, three hundred thousand or more that wander free, trample the land, and suck up the water holes. (By one account, they also turn on the faucets of farm wells, then neglect to turn them off.) They were introduced in the 1800s to serve as pack animals; now Australia has so many camels—the largest wild population in the world—that they are sold to Saudi Arabia.

The unknown, it turns out, is no place to settle down. However much the human mind may covet the concept of unfettered wilderness, it is soon exhausted by the real thing and sets about generating more comfortably familiar scenery—to eat, to watch, or simply for company. The wilderness is wild once, until you've seen it. Then it's just cold and lonely and cries out for improvement. "A surprising number of introductions appear to have been performed for what today would be considered trivial motives," a contemporary ecologist writes. "Among these, nostalgia of displaced peoples for familiar fauna to surround them would seem to rank fairly highly."

One of the most influential nostalgia movements on the American landscape was a nineteenth-century organization called the American Acclimatization Society. Although based in New York, the Acclimatization Society had branches in most every major city from New Haven to Cleveland to San Francisco. The beneficiaries of acclimatization were the societies' participants, most of whom were recent European immigrants; their goal was to introduce birds and animals from abroad. In the 1860s, Andrew Erkenbrecher, the German-born founder of the Cincinnati Zoological Garden, spent five thousand dollars importing songbirds from Europe on the premise that the songs America's birds were singing were less than inspiring. "It may be expected that the ennobling influence of the song of birds will be felt by the inhabitants," he wrote, "as well as enliven our parks, woods and meadows, which in comparison with European countries are so bare of feathered songsters."

Among the first introduced singers to succeed in the United States was the English sparrow. The first eight pairs were brought from England in 1850 and released in New York City by Nicholas Pike, a board member of the Brooklyn Institute. These failed to establish, so a second attempt was made in 1852. "Fifty sparrows were let loose at the Narrows, according to the instructions, and the rest on arrival were placed in the tower of Greenwood Cemetery chapel. They did not fare well and were removed to the house of Mr. John Hooper, a member of the committee, for care during the winter." The sparrows were set free in the cemetery the following spring; they thrived and spread. Soon, similar efforts were under way across the country, as well as in South America, Australia, New Zealand, and the then-independent island nation of Hawaii. By the 1880s, when it became evident that the sparrows were eating the buds and blossoms of fruit trees and thriving at the expense of wheat growers, it was too late; local newspapers consoled readers with recipes for sparrow potpie. The success of acclimatization is measurable today by its transparency. Like most Americans, I grew up with sparrows at the bird feeder. For the longest time I had no idea what they were; they were small and brown and flitted about, as common as squirrels or pigeons, so ubiqitous they were invisible. Even when I learned that they were English house sparrows, it took a while to register: they came from England.

Pratt remarked that a similar movement had taken place in Hawaii. We were still in the dry forest, making our way through an open area filled with young trees that had grown in since the 1960s, when park managers fenced it off and shot the last of the feral goats that had grazed there for decades. In Hawaii, the local acclimatization society was named Hui Manu—"Bird Group" in Hawaiian, although few of the society's members actually spoke the language, Pratt said. (This, despite the fact that early in the nineteenth century, American missionaries had simplified the Hawaiian alphabet to just twelve letters and a liberal use of apostrophes.) The mainland societies had aimed to transform the new landscape of birds into a refreshingly familiar Old World one—to turn the unknown into the known. The Hawaiian group charged itself with the opposite task: turning what was becoming an increasingly unmysterious and homogenous landscape into one newly exotic and unknown.

By the late nineteenth century, for reasons that scientists would not decipher for some years to come, Hawaii's native birds were rapidly disappearing. No longer did the nukupu'u visit feeders in Oahu; no longer could a collector shoot a dozen o'o—a regal black bird with long tail feathers and saffron epaulets—in a single afternoon. Hawaii, that icon of exotica, was becoming a little too familiar and quiet. Hui Manu was already busy releasing mainland game birds into Hawaiian forests: quail, grouse, turkeys, partridge, pheasant, all birds that Pratt and Foote still see from time to time in the national park today. Now the group set out to refresh the palettes of color and sound. Japanese bush warblers and white-eyes were set free, nightingales and yellowfern canaries, saffron finches, lavender waxbills, parakeets, cockatoos, mockingbirds, and magpies. They had fooled me. "It was mostly people from Honolulu that wanted to see these beautiful birds flying around," Pratt said. Many of the releases, like the three flamingos imported from Cuba by one ornithological optimist, failed to take hold. Others quickly settled in. The mynah, released in 1865 to control the cutworms and armyworms that had begun plaguing sugar-cane plantations, established itself so readily that guests of the Royal Hawaiian Hotel complained of the squawking in the banyan trees at dusk. Hotel managers concocted several schemes to scare the birds off, even playing back the distress call of the mynah, a tactic that succeeded only in making the guests' dentures quiver with high-frequency vibrations.

A more serious problem soon became apparent: the mynah was dispersing the seeds of lantana, a decorative shrub that had been imported into Hawaii in 1858 and was now spreading up into the foothills, growing in dense stands, overtaking native shrubs, and proving toxic to certain valuable nonnative creatures called cattle. Lantana today is one of more than four hundred nonindigenous plants in Hawai'i Volcanoes National Park; a few dozen pose a sufficient threat that park managers will spend millions of dollars in the coming years trying to eliminate them or contain their spread. Pratt has written a book about them. The protagonists include banana poka, a showy vine that strangles trees in light-hogging mats, like a colorful kudzu; it is the main food of the introduced kalij pheasant. Himalayan raspberry, Florida blackberry, and thimbleberry, which form impassable thickets of briars in the forests, are likewise thought to be spread by alien birds. *Miconia calvescens*, an ornamental tree from the Azores, escaped from a nursery near Hilo and is slowly making its way toward the park; in Tahiti, where it invaded in 1937, *Miconia* now comprises 80 percent of the forest.

"That's what I mean by homogenization," Pratt said. "People have brought this suite of plants and animals with them to Hawaii, and they've done the same thing everywhere else in the world. Look at the worst extreme; look at the lowlands. You can sit in the middle of Hilo, and you could just as easily be in any of twenty-five other towns in tropical or subtropical areas all over the world. Looking around you, you wouldn't know: Where are you?

"A lot of people have a difficult time understanding why we're so interested in native plants and animals. We had a group of kids from Great Britain here a few years ago working on various resource management projects, mostly removing alien plants. They worked very hard, but they couldn't really understand what we were doing. 'Why are you worried about these plants? Humans brought them here; humans are natural; therefore the invasion of alien plants is natural.' I tried to explain what our viewpoint was. Afterward, I came to feel that they couldn't understand it, because they'd grown up in a situation in Great Britain where everything they looked at had been changed by humans thousands of years ago. We have schoolchildren from Hilo visiting the park. If you ask them, 'What's a native bird?' they'll say mynah or cardinal—that's a native bird to them."

Once, in a museum, I saw the rings of a tree fossilized in a sheet of limestone: an entire lifetime fixed on the thinnest slice of a far longer, inanimate one. Perhaps the mind works the same way. The moments of the present come and go between blinks, one by one falling behind the eye, accumulating in the brain like chalk in the seabed. Only much later, in the tracing of deposited layers, does the experience of nature acquire a discernible shape: these are the shells I saw on the beach as a boy, this is the tree I climbed in, this is the bird I heard at the close of day. The stars were brighter then, there were fewer deer, it snowed more in winter. How deep do the strata run? How much time does the mind contain? How long before the memories within are reshaped by the added weight of new sediment—before the bottommost and oldest memories buckle, dissolve, and re-form, and one can no longer distinguish what really happened from the way one remembers it? That is perhaps the most vexing challenge posed by alien species: how to delineate natural history from the eye that perceives it. An inspector for the Hawaii Department of Agriculture ruefully put it to me this way: "Nature is defined by human memory, which is infinitely shorter than ecological memory."

7

Slowly I began to sense just how tall an order I had set for myself. I had come to Hawaii to step past the scrim of humanity and see nature as it had not been seen before, but clearly I had no idea what this should look or sound like nor how to identify it when it found me. The Hawaiian birds I'd thought were native were not: either they were imported from the mainland, itself long settled by European birds, or they were "exotic" introductions, by which I mean they were novel both to Hawaii and to me, and so were effectively indistinguishable to my eye. (I confess I am not the best at recognizing birds, but am perhaps no worse than the average bird viewer.) To see a true Hawaiian bird and know it as such, I would have to be guided to it—by a live biologist or by a picture in a biologist-written bird book. I was not going to get far in nature on my own. "To see the scarlet oak, the scarlet oak must, in a sense, be in your eye when you go forth," Thoreau wrote. "We cannot see anything until we are possessed with the idea of it, and then we can hardly see anything else." Clearly there was something in my eye. I started to wonder if I could ever get it out.

In January 1778 the residents of the islands of Kauai and Niihau saw something novel in their eyes—the appearance offshore of two floating islands sprouting masts like trees and inhabited by odd people: pale, with loose skins, smoke issuing from their mouths, and long hair trailing from tricornered heads. The visitors were friendly, however, so the feasts commenced and trade was exchanged; the chief of the smoking flap-skins was Captain James Cook. After a few weeks, Cook and his islands floated away again and the onshore viewers went back to their daily lives, entirely unaware that they now lived on the Sandwich Islands, so named in honor of someone called the Earl of Sandwich.

The following January, after several months charting the coast of the Pacific Northwest, Cook returned to Hawaii, setting anchor in Kealakekua Bay, on the *kona* side of the Big Island, fifty miles north of Ka Lae. One weekend, after too many rainy days in Volcano, I drove up to see it. Today Kealakekua Bay receives state protection as a marine-life sanctuary. Though narrow streets of modern homes have sprung up in front of it, the bay itself remains blue and serene, a prime spot for snorkeling if the prospect of sharks doesn't trouble you. I visited Pu'uhonua o Honaunau, a national historic park that sits on a promontory at the bay's southern end, and contemplated various religious relics and ruins and a reconstructed thatch temple. Across the bay, in a glade of coconut palms on a small promontory, I found a small white obelisk bearing a bronze plaque: NEAR THIS SPOT CAPT. JAMES COOK WAS KILLED FEBRUARY 14, 1779.

Cook and his crew spent weeks here, repairing the ships, stocking up, trading utensils and bits of iron for yams and turtles and pigs—not the large, feral pigs that Cook had released on his first stop in Hawaii, but a small homebound variety that Polynesian voyagers had carried along on their migrations for centuries. Cook wrote in his log: "We again found ourselves in the land of plenty." Wherever he went, he noted, he was received with a reverence "approaching to adoration": prostrate hordes, incomprehensible rituals. As timing had it, Cook had landed in the midst of a Hawaiian ceremony celebrating the long-awaited arrival of the god Lono—a name they now applied to Cook. The Hawaiians honored him as a great chief, or they thought he was a god, or Cook believed the Hawaiians thought he was a god—or perhaps, as some anthropologists have contended, it was only afterward in the minds of imperialist historians and missionary notetakers that Hawaiians thought Cook was a god. Whoever everybody was, everyone thought they were someone else.

God or not, Cook sailed from Kealakekua Bay on February 4. A storm rose up, a foremast snapped, and he was soon forced to return. The reception this time was notably sour, for reasons still debated: the Hawaiians' stores were spent and their hospitality strained; or their worldview had been shattered (according to the ritual, Lono was supposed to sail away and not return); or Cook now looked too bedraggled to be divine; or his behavior—erratic and vengeful, strained by thirty near-continuous years at sea—sparked ill feeling against him. Natives stole trinkets, blacksmith tongs, and a dinghy from the ship; they were pursued and fired upon; a chief was accidentally killed. Onshore, Cook found himself surrounded by an angry mob. Knives were drawn, and as one crew member later recalled, "a fellow gave him a blow on the head with a large Club and he was seen alive no more."

Afterward, the locals dismembered and burned Cook's corpse, salted his hands, paraded his bones around the island, and then divided them among the chiefs. They placed his heart in a tree, where, one rumor had it, it was eaten by a small child. It was a tragedy of miscommunication, or it was the primal act of primitive, ritually bound minds. Or, to more than one missionary, it was a visitation of God, the divine retribution for Cook's sin of welcoming false adoration: "How vain, rebellious, and at the same time contemptible for a worm to presume to receive homage and sacrifices from the stupid and polluted worshippers of demons."

What else would a traveler think, so far from home, flogged by the sea and half starved on salt rations? Brown bodies swam around Cook's ships like shoals of fish, offering strange fruits and naked adulation. It was a sort of paradise, if not quite a peaceable kingdom: these Sandwichers were constantly at war, the chief of this island sending his canoes against the chief of that island, back and forth for centuries, until 1795, when Kamehameha defeated most of them and proclaimed himself king. And the native inhabitants numbered in the hundreds of thousands, their taro plantations carved from the forest and covering the lowlands like a patchwork blanket, the islands settled more densely and uniformly even than they are today. If this was Eden, it could not last, for the same reason that no Eden ever lasts: we all hurry to see it, only to discover that it departed immediately with our arrival. Always there is a worm in the apple, and the worm is us.

By the nineteenth century Hawaii was squarely on the maps, and it was on the way to everywhere everyone wanted to go. Traders found a midpoint between East and West; whalers paused in their pursuit of whales; Christian missionaries were drawn by the evil of nakedness. Desires were exchanged. Visitors bought molasses, pumpkins, Irish potatoes, coffee, bananas, oranges, cabbages, pineapples, leather, melons, hogs, yams, sugarcane, taro, plantains, sandalwood, salt. Hawaiians bought nails, spoons, guns, cannons, ammunition. The newcomers feasted; the natives acquired everything but resistance.

Manifest destiny had begun to spill off the California coast out into the Pacific. In 1871 the number of foreigners in Hawaii was "between 5,000 and 6,000, two-thirds of whom are from the United States, and they own a disproportionate share of wealth," the U.S. minister to Hawaii wrote. "The foreigners are creeping in among the natives, getting their largest and best lands, water privileges, building lots, etc.," a Honolulu wholesaler remarked. "The Lord seems to be allowing such things to take place that the Islands may gradually pass into other hands." Alongside the old kings, a new one arose: sugar. By the turn of the century, more than a hundred thousand plantation laborers—Chinese, Japanese, Portuguese and Philippine—had immigrated to perform fieldwork that Hawaiians no longer could, because they were so few. Between 1778 and 1900 the native Hawaiian population dropped from at least three hundred thousand—and perhaps far more—to just twenty-nine thousand. Some had chosen to leave, enticed by income and the appeal of distant shores on newly drawn world maps. In 1850 alone, four thousand Hawaiians—one-eighth of the total native population at the time—embarked on whaling vessels, most never to return. "We have heard that there is no port in this ocean untrodden by Hawaiians," the minister of the interior was told, "and they are also in Nantucket, New Bedford, Sag Harbor, New London, and other places in the United States." Many Hawaiians were carried off by different newcomers: tuberculosis, whooping cough, measles, mumps, cholera, influenza, smallpox, the common cold. A proverb came into being: *Lawe li'ili'i ka make a ka Hawai'i, lawe nui ka make a ka haole.* "Native death takes a few at a time, the foreigners' death takes many."

Hawaii's birds were quietly suffering a similar fate. They had been fading for some time already: in recent years archaeologists have discovered that well before the arrival of Captain Cook and the Europeans, native Hawaiians had hunted at least seven species of flightless goose and two species of flightless ibis to extinction; and several species of honeycreeper were barely hanging on when the rest of the world began stopping in. Increasingly, the remaining lowland forest habitat was replaced by sugar and pineapple plantations. By night, European rats stole eggs from nests; by day, the mongoose, an Indonesian introduction, did the same.

But the most fearsome enemy was disease. In 1893, after ten years of surveying the fauna of Hawaii, the biologist Robert Perkins noted that many of the birds on Oahu and the Big Island had swellings on their legs and feet, and in some cases were missing one or more claws or parts of toes. He sent specimens to a lab in Washington, and the analysis indicated the work of bumblefoot, or bird pox, a degenerative disease common among chickens, turkeys, pigeons, and other birds but entirely unfamiliar to the unworldly forest birds of Hawaii, which succumbed in countless numbers. Although the pox arrived with nonnative birds, its transmission requires mosquitoes. These had arrived several decades earlier when the crew of a Central American ship emptied their casks of stagnant water, and the larvae living in it, into a stream on Maui. The mosquito became a ready vector for bird pox and, subsequently, avian malaria. The parasite responsible for avian malaria, *Plasmodium relictum,* probably arrived in Hawaii at the turn of the century aboard one of the many foreign birds then being released by acclimatization groups. By 1930, an epidemic of avian malaria had swept through the forest. Hawaii's birds became paradigms of island vulnerability and immunity breakdown, so much so that in the early 1980s, when biologists on Guam began asking why that island's birds were disappearing, many found it impossible to believe that anything but disease could be responsible.

The malaria pandemic continues today among Hawaii's native birds. Its front line runs straight through the upland forests of Hawai'i Volcanoes National Park, through the cluster of drab brown buildings that comprises the park headquarters, and across the institutional steel desk of a parasitologist named Carter Atkinson. "There was a major epidemic at the end of 1992," he said one afternoon. "We found birds too weak to fly. We brought them in but couldn't do anything." Malaria works by destroying the red blood cells. In normal blood, the cellular bits make up roughly half of what a scientist sees. In the birds, that portion was more like 10 to 15 percent. "Their blood was like water."

Atkinson is slight, with receding hair and wire-rim spectacles. Although invariably friendly, he radiates an awkwardness that suggests he would be more comfortable by himself, doing pretty much anything other than talking. If he were a bird, he might be a night heron or a small saw-whet owl: liable to stand there blinking in the illumination of a flashlight, then dart away the moment the light roamed off. Atkinson first came to Hawaii in 1977 to help with a statewide survey of forest birds, and in 1991 he moved there. His self-appointed task has been to catch and count mosquitoes, to figure out how many there are and where they breed, and to gather and to monitor the avian-malaria epidemic. The basic method of catching mosquitoes involves a contraption called an ovipositor trap. Atkinson had one on hand and lifted it onto his desk: a plastic tube with a motorized fan at the top end and a pan of stagnant water at the bottom. The trap is left out overnight in the forest; adult mosquitoes lay their eggs on the surface of the water and are then sucked up the tube into a collection bag. One thing Atkinson has found is surprisingly few mosquitoes. "We're lucky to get five or ten in a night," he said. "If we hadn't actually gone out and found dead birds dropping out of trees, I would find it hard to believe that a population of mosquitoes this small could support an epidemic."

The fatal difference lies in Hawaii's native birds, which are far less resistant to *Plasmodium,* the malaria parasite the mosquitoes carry, than are the introduced birds. In the 1980s, Charles van Riper, an ornithologist from the University of California at Davis, conducted an experiment in which he injected different bird species on the Big Island—the native apapane, i'iwi, and amakihi, and the introduced Japanese white-eye and red-billed leiothrix—with the *Plasmodium* parasite. The results were eye-opening: every single introduced bird survived infection, whereas only 42 percent of the honeycreepers did. Through his own experiments, Carter Atkinson found that up to 50 percent of the mosquitoes that are trapped after they've drawn blood from a native bird carry malaria; that is, as many as half the native birds with blood to give carry the *Plasmodium* parasite in their blood. And the native birds maintain infective stages of the malaria parasite in their blood for much longer—up to thirty days after being bitten—than introduced birds do.

If I was having trouble seeing Hawaii's birds, evidently part of the reason was that few of them were alive and available for viewing, at least in the areas I had visited thus far. The conflict between birds, mosquitoes, and the malarial parasite plays out on a landscape that stretches from the coastline up to forty-five hundred feet in elevation, high enough to include Kilauea Crater and Atkinson's office, where I stood. Above that altitude, the cold nights prevent the mosquitoes from breeding. The forest birds—the native ones, at least—are hardly found in significant numbers below the forty-five-hundred-foot mark; the upper elevations are their last refuge. Most birds infected with malaria are found between three thousand and forty-five hundred feet—precisely where the range of vector and host overlap. In the 1980s, van Riper found that the native bird with the highest rate of infection was the apapane—probably, he surmised, because it regularly flies from the upper elevations, where it nests, to the lower elevations, into the malaria zone, to collect nectar. The i'iwi shares the apapane's nomadic life and infection rate, yet once infected, the i'iwi is far less likely to survive. Atkinson has found that the malarial parasite kills 60 percent of the apapanes and amakihis that become infected, and close to 90 percent of the i'iwis. By process of elimination, the native birds most resistant to malaria are the ones most prevalent today. The i'iwi, once one of the most abundant and widely distributed of all the honeycreepers, is far less numerous today. In contrast, the apapane and amakihi, now two of the most common honeycreepers, have even managed to reinhabit some lower-elevation forests they had previously ceded to the mosquito.

Atkinson mentioned that he had a few honeycreepers in his lab, if I was interested in seeing some. He took me around to a small brown cabin set apart from the other park administration buildings. Atkinson's experiments involve exposing birds to malaria-bearing mosquitoes and closely studying the consequences. To avoid compromising the experiments, any human entering the lab must obey a few basic quarantine measures. Atkinson led me through the cabin's front door into a narrow foyer, where we removed our shoes and put on plastic slippers. Then we passed through a second door and into a narrow hallway with five doors labeled A through E, each with a window. In room D, I could see a series of screened cages sitting on a benchtop, aflurry with breeding mosquitoes. Atkinson had developed an elaborate experimental protocol to precisely control the amount of malarial parasite that each bird receives. This involved, among other things, drawing blood (and with it the *Plasmodium* parasite) from an infected honeycreeper; injecting that into a canary, in which the parasite would reproduce for ten days; "exsanguinating" the canary (draining and collecting its blood) and reisolating the now-prospering parasites; injecting those into ducklings; exposing the ducklings to mosquitoes to permit the latter to load up on parasites; and finally, exposing these mosquitoes, visible in Room D, one-on-one to honeycreepers, infecting them.

Room B was where Atkinson maintained the native forest birds between experiments. This room was not open to visitors, but from the hallway I could look through a one-way window into a small chamber where a dozen birds—mostly sparrow-size and dull lime in color, except for one larger, bright red one—flitted around. For company they had four potted o'hia trees with quarter slices of orange stuck on several branches, recalling a sort of Charlie Brown citrus tree. Peering in, I felt a bit like a parasite myself. Atkinson said, "There are twelve birds in there: eleven ama-kihis, one apapane. Half now have malaria." He turned a knob on the wall, activating a ceiling sprinkler in the bird room. "We have a mister that really gets them turned on. As soon as you switch it on, they start to sing." On this occasion they did not sing. Atkinson showed me the rest of the facility, and then we returned to his office.

He is still amazed and alarmed by the high mortality among i'iwis exposed to avian malaria. In one experiment, every single i'iwi, even those bitten only once, developed malaria within the first four days. After thirty-seven days, all but one were dead. Along the way, their food intake fell sharply, though by how much exactly Atkinson couldn't say—the birds had died so fast he had stopped keeping track. Their weights, on average, had dropped by 13 percent, the lightest succumbing first. In the end only a single i'iwi—a male, and the heaviest of the lot—survived. When Atkinson exposed it to another round of infection five months later, it continued to thrive. He shook his head just thinking about it: "They were so susceptible." The experiment led him to surmise that young i'iwis are especially susceptible to malaria: they weigh less and are subordinate to adults and other birds at feeding time. Only a small fraction of juvenile i'iwis are likely able to recover from malaria and develop resistance to reinfection, he concluded.

What makes the i'iwi so much more susceptible to malaria than native birds like the amakihi and apapane? With Sue Jarvi, a genetics researcher at the national park, Atkinson has begun to explore how the birds' genetic makeup contributes to their differences in vulnerability. Elsewhere on the Big Island, ornithologists have established a captive-breeding program for several species of Hawaiian birds. By rearing native birds in captivity and releasing them into the wild, biologists hope to sustain the otherwise dwindling wild populations. Jarvi's and Atkinson's research could prove useful to that endeavor. Atkinson said, "The long-term rationale is, if you can identify birds with a certain genetic background that are more resistant, then you could incorporate that into a captive-breeding program. If you're going to rear birds and release them to another island or to a new habitat, you could start with ones that are disease resistant." The challenge of pinpointing a gene or a suite of genes somewhere in the deoxyribonucleic archipelago that might account for the i'iwi's high mortality rate promises to be painstaking. Jarvi's task is made slightly easier by the unique pedigree of the honeycreepers. Genetically speaking, the *Drepanididae* are a tight-knit family: the individual species are all descended from a single ancestral line, and they evolved in isolation, rarely mixing or hybridizing with non-honeycreepers. Consequently the DNA of one honeycreeper species is virtually identical to the DNA of another—identical enough, anyway, that whatever genetic differences separate them should stand out clearly to a trained eye.

"It might be that the more common species will develop resistance in the long run, say a hundred years," Atkinson said. "But the birds with low populations, like the i'iwi, are in the most trouble. An epidemic that wipes out ninety percent of their population could be enough to push them over the edge. The saving grace is that the endangered birds live at high elevations, where the current crop of mosquitoes can't live. What would really mess things up is if somebody introduced a temperate mosquito, one that did well at high elevations. That would be disastrous."

The interplay between disease, host, and vector is dynamic, constantly evolving, and turning in new and unexpected directions. Just as the human ear, when assaulted by some noxious and continuous noise, manages over time to occlude it, to proceed around it as if it were hardly more than the buzzing of a fly, so too the behavior of the native birds has altered under the oppressive rule of malaria. Back in 1968, while studying avian diseases in Hawaii, the biologist Richard Warner noticed that all the nonnative birds in his experimental cages slept with their bills and faces tucked into the fluffed feathers on their backs and with their legs tucked underneath them. In contrast, the honeycreepers slept whatever way, with everything exposed—a habit, Warner speculated, that would make them more likely to be bitten by malarial mosquitoes. When Charles van Riper looked at the behavior in 1986, he saw something quite different: all the honeycreepers he saw slept with their heads tucked and one leg raised into their feathers. Evidently, in less than twenty years, a behavioral shift had taken place. It is not the honeycreepers learned a new sleeping position. Rather, the old habit was selected against: the birds who exhibited it were eliminated, while those few who slept defensively were able to live, breed, and flourish.

Van Riper found a comparable shift in the honeycreepers' eating habits. The nectar-producing trees in Hawaii bloom along a gradient: the trees in the lowlands reach the peak of flowering in the summer and fall, whereas the high-elevation trees reach peak bloom in the winter, the slope of a mountainside rippling up and down with color and fragrance as the year cycles around. To gather the most nectar, then, a honeycreeper must pursue these ripples, like a surfer migrating with the seasonal swells. It moves uphill in the winter and downhill, into the malaria zone, in the fall—precisely the season, it so happens, when the mosquito population is blooming. No wonder, van Riper concluded, that i'iwis are infected in greatest numbers during the fall season. Except that the i'iwis do not migrate seasonally up and down the slope, at least not anymore. Rather, van Riper found, the i'iwis, the apapanes, and a few other native birds make this commute daily. They leave their perches early in the morning and begin working their way downhill, reaching the lowest elevation just as the *Culex* mosquito, which has been up all night in search of blood, is at the nadir of activity. At dusk, the birds gather and fly home; by eight in the evening, when *Culex* is again mustering its nighttime forces, the birds are asleep uphill, safely out of range. At one time the birds may have migrated seasonally, but selection pressure clipped that habit short. Van Riper concluded: "What was once probably a gradual movement of the birds downslope following the flowering of nectar-producing trees, has now evolved into a daily circular pattern."

Natural selection has begun its pruning, favoring native birds with habits that, however unwittingly, protect them from introduced mosquitoes. (One might fairly ask: if the selective pressure is applied by an alien species, is it still "natural" selection?) Yet selection is egalitarian: what applies to the host applies equally to the parasite. At the start of the twentieth century, the primary reservoirs of *Plasmodium relictum* were non-native bird species. Birds like the California quail and the red-billed leiothrix brought it in, worked their way into the forests, and served as hosts while the mosquitoes dished out the parasite to the honeycreepers and other Hawaiian birds. The situation has since evolved. In the early 1980s van Riper found that the leiothrixes, the Japanese white-eyes, and other introduced birds carried only low levels of the parasite in their blood; a decade later, Atkinson could find only a few nonnative birds with detectable infections. The parasite has switched primary hosts, from the introduced birds to the native ones. The nonnative birds no longer play the Trojan horse in the malaria epidemic; now the native birds are the vessels of their own doom. Although one might surmise that the introduced birds have become more resistant to the malaria parasite, that is not what Atkinson thinks; rather, he believes that the parasite, to its own advantage, has become weaker. "If a parasite is really virulent, it kills its host so fast that it can't be transmitted—the parasite dies out. So you're gonna see selection for strains of the parasite that the native birds can tolerate long enough for the mosquito to pass it on to something else." Recently van Riper has found some experimental evidence to suggest that the Hawaiian strains of *Plasmodium relictum* may be less virulent than their mainland counterparts.

Atkinson has developed a hypothesis. As he sees it, a virulent strain of avian malaria reached Hawaii a century ago and wiped out all the native birds in the lowlands. Then, over time, a less virulent, intermediate-strength strain of the virus was selected. It still kills a high percentage of native birds like the i'iwi, but other native birds like the apapane and amakihi can tolerate it; the introduced birds, more resistant to begin with, are effectively rendered immune. "You still have a high level of mortality, but there are enough native birds around that they can maintain the cycle." It is something of a paradox: the same conditions that fostered the evolution of the honeycreepers—genetic isolation, small populations—may be driving some sort of adaptation, a coevolutionary compromise, between vulnerable native species and their foreign parasitic enemy. Not that Atkinson admits to optimism. The tides of evolution are murky and shifting, not to be counted on, especially where viruses and bacteria are concerned.

"It's important to remember that the parasite is also evolving and changing. And we have no idea how many different strains or types of parasites are out there. It's something that people kind of forget about. Nobody's looking at that at all."

8

"The fate of remote islands is rather melancholy," Charles Elton remarked in *The Ecology of Invasions by Animals and Plants.* "The reconstitution of their vegetation and fauna into a balanced network of species will take a great many years. So far, no one has even tried to visualize what the end will be. What is the full ecosystem on a place like Guam or Kauai or Easter Island? How many species can get along together in one place? What is the nature of the balance amongst them?"

Elton was well into his career when *The Ecology of Invasions* appeared, in 1958. His first book, *Animal Ecology,* which he'd managed to write in just eighty-five days in 1926, at the age of twenty-six, established him as a leading voice in the new science then beginning to congeal around the word *ecology.* The term was coined in the nineteenth century by the German biologist Ernst Haeckel, who defined it alternately as "the sociology and economy of animals" and "the study of all the complex interrelationships referred to by Darwin as the conditions of the struggle for existence." Literally, the term refers to the study (*logos*) of the immediate surroundings (*oikos*)—effectively, what goes on in our backyards.

Today the science of ecology is generally understood as the study of communities in nature. An ecologist tries to detect patterns in these communities and develop explanations for them—to ask not only "what" (the purview, traditionally, of the natural historian) but also "how" and "why." The foremost questions are as simple to state as they are difficult to answer: How do ecological communities form? How does an individual species fit into an ecosystem? What sort of structure binds these various organisms into an assembled whole? Today's invasion scientist asks a similar array of questions: Why do the vast majority of invasions fail to take hold in their new environments? What is it—in the invader, in the recipient ecosystem, or in some combination of the two—that keeps a newcomer from settling in? Why does a given group repel, or fail to repel, an arriving individual? "A central task," a contemporary ecologist summarizes, "is to explain why certain groups of species exist together in time and space. A satisfying answer must answer both why certain species successfully coexist and why they are not joined by other species."

In short: How does nature work? How should one visualize it? Is it a finely tuned machine, like cogworks or the insides of a watch, liable to grind to a halt if too many loose screws are tossed in? Or is an ecosystem instead like an airplane: remove some critical rivets—the native species integral to its structure—and the entire infrastructure crashes to the ground? Some scientists refer to these rivets as "keystone species," evoking less an aircraft than a vaulted cathedral. Or perhaps an economic analogy is more apt: an ecological free market of producers and consumers, all competing for limited natural resources, all buying, stealing, or otherwise exchanging the nutritional equivalent of energy vouchers. Is it a machine, an edifice, an organism, a community-watch program, an international bank? To understand nature—much less preserve and protect it—one must conjure the right metaphor for it. If nature is like a clockwork, the tools of veterinary medicine don't apply. If nature is a cathedral, no airplane technician can fix it. For Elton, the study of invasions offered a way to probe and test nature's essential shape and operation. Anyone who journeys down this path today, across the shifting landscape wrought by introduced species, hoping to see nature for what it truly is, sooner or later finds himself in Elton's guiding company, sizing up the natural world through his eyes.

I have a photograph, taken in the 1930s, of the young Elton on a motorcycle, riding down a road near Oxford, England. A large, stuffed sack hangs on his back wheel like a saddlebag, as if Elton is embarking on a long and arduous trip. I like to imagine him as ecology's Che Guevara, fond of daring and high speeds, except that he is wearing a tweed coat and was by all accounts a shy and unassuming man. When a former student first met Elton in 1960, he mistook the biologist—by then the distinguished head of the Bureau of Animal Population, which he founded in 1932—for the janitor. Early on, Elton was an amateur boxer, and he liked to say that one of the good things about walking down a busy street in Oxford was that it required him to dodge oncoming people and thereby improved his footwork. He conducted fieldwork on some Arctic islands and later visited South America, but on the whole he avoided travel, preferring instead the glades and fields of nearby Whytham Woods. He devoted the early part of his career to capturing and counting rodents—in 1942 he published a book called *Voles, Mice, and Lemmings*—in an effort to understand how and why the populations of certain animals fluctuate over time. Insofar as his professional world involved a great many mousetraps, the sack on the back of his motorcycle in that photograph could be said to contain his worldly possessions.

Let us place ourselves in his seat, then, as he rides through the Oxford woods and envision the natural landscape as it presents itself to him. Placid from a distance, this landscape in fact roiled with philosophical tensions about how it should best be viewed. Pulling from one side were the proponents of order, integration, stability. They saw an ecological community as a sort of limited-membership club, its constituents organized and bound by time-honored alliances and dependencies that, left undisturbed, could effectively bar new entrants from settling in. Species were where they were, and not elsewhere, in part because certain communities would not have them. At its extreme, this view considered an ecosystem literally as an organism—a superorganism—able to discern member from nonmember, native from alien, self from nonself.

Meanwhile, pulling from the other side were the advocates of individualism and happenstance. The physical environment varies through time and geographic space; every species, and each individual, can tolerate a range of conditions. The reason a species is here and not there is no more than serendipity: a random act of dispersal that happened to drop an organism—a windblown finch, a ballooning spider—into a physical environment it could withstand. No overarching principles, no community-wide ecological laws. "Every species of plant is a law unto itself," declared New York Botanical Garden biologist Henry Gleason, the main (and for a long time, only) proponent of this view; an ecological community is "not an organism, scarcely even a vegetational unit, but merely a coincidence." Is the whole greater than the sum of its parts—or is it just some parts in the same place at the same time? Are there rules—or instead, as Gleason contended, are the rules "merely abstract extrapolations of the ecologist's mind"? Is there a structure to nature, or is it in the scientists' heads?

These two poles marked a growing divide between plant and animal ecology. By 1927, when Elton's *Animal Ecology* appeared, plant ecologists already had spent years carving out a sizable intellectual territory for themselves. Central to their outlook was the notion of succession, the apparent trajectory of certain plant communities, particularly forests: the early years of mad rush, of quick-sprouting seeds and short-lived, colonizing weeds; then a gradual maturing into a climax community of trees that, barring windstorms and handsaws, could persist for decades or even centuries—relatively unchanged, apparently stable, and, it seemed to their studiers, largely closed to incursions by nonmember species. The struggle for existence continued quietly, everywhere and throughout—limbs and leaves stretched for the sun's attention, roots probed for buried nitrogen treasure—but now took on an added, community-wide dimension. As one researcher of the period phrased it, "Each species competes with those around it and in this competition the individuals might be said to stand shoulder to shoulder against the common foe."

The animal ecologists, meanwhile, were still figuring out where on this property to erect their tent. Animals were proving to be more vexing ecological research subjects than plants. For a start, they rarely stood still; consequently, the boundaries of their communities could be impossibly vast or difficult for a researcher to traverse. (Try following bison on the Great Plains, or a snake in the forest, or a cockroach in your kitchen.) In addition, these boundaries often enough bore little or no correspondence with physical real estate. (How does one delimit a community of starlings?) Elton, in *Animal Ecology*, was frank: "The writer has found that it is almost impossible to make even a superficial study of succession in any large and complicated community, owing to the appalling amount of mere collecting which is required, and the trouble of getting the collected material identified. When one has to include seasonal changes throughout the year as well, the work becomes first of all disheartening, then terrific, and finally impossible." Studies of animal succession, he advised, were best carried out in simple, confined communities: decaying logs, rotting animal carcasses, small, brackish pools, piles of dung. By 1954, the underlying philosophical landscape had solidified somewhat, but not yet enough to please Elton. "Animal ecology is still in such an embryonic stage of thought (though it still has all too many facts to swim around in) that it can hardly be said yet to have completed its neural fold. The facts are rather chaotic; many of them are not facts at all; its theories are poised uneasily between arm-chair pipe-dreams (valuable as models of thought) and ready-to-wear mathematical models that fit badly and are already bursting at their seams." At times Elton avoided the term *ecologist* altogether and described himself simply as a zoologist.

The issue was precisely that: How to consider the individual (an organism, or a species—the units of zoology) in relation to its group (ecology). How do disparate parts become a whole? Darwin got as far as figuring out the origin of new parts—the species. Every organism varies slightly from the next in its ability to survive and reproduce; it is tossed onto an Earth of limited capacity, into a pool of limited food, mates, and refuge; and the game of natural selection begins. The winners survive to reproduce another day, spawning a new generation, ever so slightly different from the parent, that continues in the struggle; the losers die without sufficient progeny and so leave no record of themselves and their evolutionary inadequacy. In spelling this out, however, Darwin found himself faced with a paradox that, despite five revisions of *On the Origin of Species*, he could never quite resolve. The world of the individual, and of the individual species, is nasty, brutish, and short—a tooth-and-nail "war for existence" fought, essentially, over crumbs, in a timescale best described as fleeting. Yet the world of the aggregate—the coral reef, the tropical rain forest, the mountain lake—is remarkably quiescent and persistent, seemingly stable over the long term, an entrenched and "entangled bank" of species on the whole. Somehow, a drawerful of short, fraying, often mutually repellent strands of fabric meld to form a rich and lasting tapestry. How does each thread relate to the whole? What role, if any, does the individual play in sustaining the larger scheme that surrounds him? What is its value in nature? Those questions lie at the heart of modern ecology and conservation. Although the arguments and evidence have grown more sophisticated and subtle since Elton's time, the ancient poles still magnetize the scientific landscape, pulling researchers this way and that like iron filings, and lining them up one against the other: the community-assemblyists versus the individualists; the holists versus the reductionists; integrationists versus disequilibriumists; determinists versus null-theorists; believers versus atheists.

Which road to follow? From *Animal Ecology* onward, Elton argued that distinct groups of plants and animals are bound together in a multifactorial, time-tested manner; they comprise not "mere assemblages," but rather are "closely knit communities or societies." Darwin had noted that even a serene meadow is a "complex web" of relationships, in which a population of red clover can be distantly affected, through intermediary ties to pollinating bees and bee-eating mice, by the number of house cats on the prowl. Elton had spent sufficient time on his motorcycle setting mousetraps and counting voles; he knew the entangled bank firsthand. But he understood, too, the limits of metaphor. He was skeptical of the clockwork simile, the common view that a natural community is like a tightly integrated system of cogs and gears that require each other for the whole to function properly. Nature rarely proceeds so mechanically, Elton countered. If one can imagine a clock in which each cogwheel runs on its own mainspring, each at its own speed, instead of all of them running from a single spring, Elton proposed, then perhaps the simile had merit. "There is also the difficulty that each wheel retains the right to arise and migrate and settle down in another clock, only to set up further trouble in its new home"—the phenomenon known as ecological invasion. If nature is indeed a clock, it is a clock with interchangeable parts.

Nature demanded a dynamic metaphor. The natural scientist had to incorporate the clear fact of invasion—that in certain cases a species could transcend the evolutionary bonds of its native habitat and work its way into a new one. At the same time, Elton argued in *The Ecology of Invasions*, one had to account for the apparent fact that certain environments, notably islands like those in Hawaii, seem especially vulnerable to invasion. Laboratory studies at the time suggested that simple populations, comprised of few species, are more prone to wild oscillations, even to outright extinction, than are larger, more complex assemblages of species. Less diversity equals lower stability; and lower stability, Elton concluded, widens the door to foreign species. "The balance of relatively simple communities of plants and animals is more easily upset than that of richer ones," he wrote, "and more vulnerable to invasions." Remote oceanic islands like Hawaii, he added, are particularly simple, and therefore particularly prone: they contain fewer native species than mainland ecosystems do; their flora and fauna are constrained to less area, and that area contains fewer sustaining resources. In effect, Elton was among the first natural scientists to articulate a link between biological diversity and ecological health. Greater diversity conveys a degree of "biotic resistance," in Elton's phrase, that helps preserve the integrity of an ecosystem over time. A natural, undisturbed community of native species could be thought of as an immunological system; invasion, its disease.

Out there, *silva rerum*, the forest of things; inside, a metaphor, a model, a map. Where does one end and the other begin?

A scientist typically engages this question in a conceptual room known as an experiment. Although *experiment* can be thought of as a verb—an activity that unfolds over a period of time—it is perhaps better imagined as a noun, an arena in which the human mind and selected elements of the natural world spend some moments together hashing things out. In practice, an experiment can take any number of forms: a mile-long particle accelerator; a tabletop vacuum chamber; an array of wire-mesh cages filled with hungry crabs and plugged into a mudflat; a series of petri dishes filled with agar and bacteria and studied for days under a microscope. Even a microscope or telescope can be thought of as a sort of experiment, constructed as it is based on certain ideas about optics and the nature of light and requiring a great deal of skill to use properly. In 1800, several decades after the microscope was invented, the renowned French pathologist Marie-Françoise-Xavier Bichat considered the device so unreliable that he banned its use in the laboratory.

Above all, an experiment is an artifice, a conceit. It is a container, its confines established by the researcher in order to study a certain parcel of nature. This is an inherent and unavoidable fact. In a perfect world, the ideal experiment would be a "natural" experiment—something readymade and stumbled upon that can be studied immediately without any further organization, subdividing, manipulation, or managing required on the part of the investigator. A series of similar islands, ponds, trees, or even rocks can offer something close to this ideal for studying the habits of certain birds, fish, snails, or slime molds. But of course no two islands, ponds, trees, or rocks are exactly alike, so any comparison between them can never be precise. Purely natural experiments are impossible to come across for the simple reason that nature, unlike the designers of shopping malls and laboratory test tubes, declines to repeat itself. By and large, experiments are designed, not discovered—made, not born. In a sense, experiments occupy the same epistemological purgatory as gardens: not entirely natural, but not entirely unnatural either, perhaps a bit like what Thoreau called "a half-cultivated field." Here is a plot of ground set aside with the express aim of generating a few small fruits of knowledge. You will require a trowel and perhaps a low fence or even a wall to keep out the weeds and woodchucks. Certainly you could forgo these tools, but in the long run you will wind up with either gnawed cabbage or impenetrable brambles, both equally difficult to digest.

Think of an experiment as a small piece of real estate fenced off from the rest of the busy world. One of the most influential experiments in ecology took place entirely on an orange. In the late 1950s Carl Huffaker, a Berkeley ecologist and the owner of the orange, set out to test long-standing ideas about what makes a community of organisms stable, or persistent over time. Thirty years earlier, a pair of mathematicians had developed several equations (the Lotka-Volterra equations, so named for their originators) to characterize how predators and their prey interact and coexist in a community. The equations describe two oscillating and slightly overlapping curves: the prey thriving and rising in number; the predator thriving (by eating the prey) and rising in number; the prey population falling due to increased predation; the predator population falling due to a drop in available prey; the prey increasing due to a drop in predation; and so on up and down, week after month after year. A persistent cycle: stability.

Very neat and elegant—except that in the laboratory, reality did not match the graph. In 1934, in a classic experiment much admired by Elton, the Russian biologist Georgyi Frantsevich Gause dropped some prey (a species of protozoan) and a predator (another protozoan) into a test tube containing a limited supply of food (oats). The prey thrived briefly, and the predators thrived briefly on the prey; but the predators quickly ate the prey to extinction and then, after a day or two of starvation, went extinct themselves. No cycle there; barely half an oscillation. Gause next tried giving the prey some oatmeal sediment to hide in, to perhaps sustain their numbers for a little longer. Again the predators quickly ate everything in sight, then went extinct—at which point the hidden prey emerged from the sediment, fed, thrived, and reproduced, and their numbers went through the roof. No cycle there either; rather, half an oscillation, one extinction, and one population explosion. In short, Gause couldn't get the system to behave like the theory. Only by adding some new predators and prey every few days or so—the final version of his experiment—could he generate the cycles of populations predicted by Lotka and Volterra.

Looking back, Carl Huffaker wondered if perhaps Gause's microcosm had been too simple. After all, in the wild, predators and prey do manage to coexist for more than five days at a stretch. For his own experiment, Huffaker employed two species of mite: *Eotetranychus*, a six-spotted mite that eats oranges, and *Typhlodromus*, which eats six-spotted mites. Huffaker cast the prey mites loose on their own edible island of orange—actually, just the top one-twentieth of the orange, which made it easier for Huffaker to monitor and count his subjects. (The rest of the orange was wrapped in paper and sealed off from the mites—terra incognita.) Eleven days later, Huffaker set loose the predators.

As it turned out, Huffaker's initial setup was also too simple. Invariably, within about a month of colonizing the orange, the predator mites ate the prey mites to extinction and then, a couple of days later, starved to death. So Huffaker remade the mite world. Now he used bunches of oranges, four at a time, arrayed in trays and linked by fine wire bridges. He covered some of each orange in paraffin or Vaseline (impassable and inedible to mites), so that the total area of orange food equaled the same surface area as before. This gave the prey mites somewhere to go to avoid their predators, and it enabled the predators to change habitats when prey became scarce.

Again the predators found the prey, ate them, and went extinct—this time taking slightly longer in doing so. Huffaker made the experiment yet more complex, adding further oranges, bridges, and barriers. By the end, he had built a veritable mite metropolis: some two hundred fifty oranges linked with an array of paper and wire bridges, wooden pegs (which permitted mites to jump from one orange to another), Vaseline barriers, and inedible but traversible rubber balls. Here, the prey mites and predator mites dwelled together for many months, through three oscillations in their respective numbers, until after two hundred forty days their final numbers failed.

Huffaker's experiment is considered a classic because it highlighed the ecological importance of refuge. If an ecosystem is big enough and patchy enough, and permits a species—predator or prey—somewhere to go in hard times, the species will survive longer. The other organisms that rely on that species likewise can survive longer, and the whole web of dynamics gains a degree of persistence. Native Hawaiian birds like the apapane and amakihi continue to elude avian malaria in part because they can live above an elevation line that the disease vector—the mosquito—can't cross. Huffaker's microcosm likewise informed one of Elton's concerns, the weakness of island ecosystems. Island species, limited in space, typically have fewer members than similar species on the mainland; purely arithmetically, a species with few members will go extinct more quickly than a species with many members. A small species on a single patch of ground, with limited refuge, is doubly vulnerable. Any catastrophe—drought, hurricane, disease, the introduction of foreign predators like avian malaria or the brown tree snake—has a greater chance of driving the species extinct in one fell swoop. In Huffaker's more complex microcosms, the mites on any particular orange not only had somewhere to go, they had somewhere to come from; their populations gained regular infusions of new members from nearby oranges. A species on a real-world island rarely gains such a benefit. The more remote it is—the farther from the mainland or another source of additional members—the lower the odds of an infusion.

That explains why island species are more vulnerable to extinction in the face of invasions. But are islands more vulnerable to invasion per se? Is there something about the structure of island ecosystems that permits easier access to an invading organism? Elton believed so, and drew his analogy from the social world of humans. As he saw it, an ecosystem is comprised of species each occupying their own niche, performing something like a job: predation, rumination, decomposition. "The 'niche' of an animal means its place in the biotic environment, its relations to food and enemies." Every day, every moment, is a Darwinian exercise in jostling and job competition, with each niche occupied by the most fit and efficient organism for the task.

The clearest experimental evidence had come in 1934 from Gause, who put two species of paramecia (both vegetarian) into a test tube and watched as one invariably beat out the other. The outcome gave rise to the notion of "competitive exclusion": two similar species cannot occupy the same niche indefinitely. The superior competitor wins. By extension, an ecological invader succeeds in its new environment by claiming an existing niche—through predation or disease, or by outcompeting a native species for the same job. This puts island ecosystems like Hawaii's at a disadvantage, Elton argued. They are younger and simpler (that is, they contain fewer species) than their mainland equivalents, so there are more available niches—more job openings for introduced species to fill. Also, being remote and small in membership, island species are not "steeled by competition"; they are unaccustomed to high-intensity elbowing and thus are ill equipped to defend against it when it arrives from the mainland. One need only note the multitude of invasions that occur in agricultural areas—habitats and biological communities "very much simplified by man," Elton wrote—to recognize the tragic link between low richness and high vulnerability.

By the 1980s the advent of powerful computers had made it possible for scientists to model and manipulate virtual ecosystems and so probe the link between species richness, stability, and "biotic resistance." To be sure, the computer models are simple systems compared with the outside world. Still, computer models enable scientists to control a wide array of cyberbiological factors and to more closely track the effects of their tweakings. In one striking experiment, the biologists Mac Post and Stuart Pimm dropped some computer organisms onto a computer island. Each species—an array of plants, herbivores, and carnivores—was described mathematically in terms of its type, size, territorial range, and food needs. One by one they entered this digital realm. Post and Pimm made no attempt to build a particular community (although they did follow a logical pattern of plants first, then herbivores, then predators). Nonetheless, communities did form. And the dynamics were striking: as the communities grew more rich in species, the harder it became to add new species. Elton seemed to have a point. "Simply," Pimm wrote, "the more species there are, the harder the system is to invade."

The test of any experiment comes in moving from local observations to general truths—in extending the microcosm to the macrocosm. To what extent does an orange, an algae-filled beaker, or a computer model accurately represent what goes on regularly in the wild? Here, nature provides a ready looking glass. "The Hawaiian Islands provide an obvious system on which to test ideas of community resistance, because they have received so many introductions," Pimm has noted. And in certain respects, the ecological reality of Hawaii seems to confirm the predictions. Looking back over the history of bird introductions in Hawaii, Pimm found that the probability that an introduction would fail increased over time in proportion to the number of introduced bird species already established. The more physically similar the birds—that is, the more likely they presumably would compete for the same niche—the higher the probability of failure. Likewise, some biologists have noted that the ecosystems of the Hawaiian Islands are essentially segregated into two categories: the high-elevation forests, which are mature in geologic age, still largely undisturbed, and rich in native species; and the lowlands, which consist largely of foreign species that have been thrown together in the past two hundred years. Most of the invaders are established in the lowlands, not the highlands—lending apparent credence to the notion that mature ecosystems have unique resistive properties denied to younger ones.

For Elton, the link between invasions, species richness, and biotic resistance was more than an intellectual puzzle. It was a key to saving places like Hawaii, where the native biological tapestry is being undone and rewoven into a commonplace design. "There is a prospect of being able to handle our biological affairs by the better planning of habitat interspersion and the building up of fairly complex plant and animal communities," he wrote in *The Ecology of Invasions*. What is at stake, for Elton then and for ecologists today, is biological diversity—the stunningly rich variety of life-forms that crawl, fly, trot, swim, drift, scuttle, or slime across Earth's surface, or sprout and blossom from it. Not merely the physical state of biodiversity—whether and where it exists, and what condition it is in—but also, and perhaps as important, its agreed value in both conversation and conservation. Does biological richness make a difference in nature? Are intact, mature, species-rich ecosystems inherently stronger than ecosystems that have been disturbed by people or heavily invaded by exotic organisms? Is a native ecosystem somehow measurably superior to a nonnative ecosystem? Is natural better than nonnatural? How much does biodiversity matter?

9

A fence, at its simplest, is a line. It is the definition of desire. On one side, there is what we want; on the other side, what we don't. What belongs, what does not belong; accepted, unacceptable. Of course, more than a line, a fence is also a barrier. It not only designates, it separates. It ensures no mixing, no mingling, no internecine traffic. A fence is not unlike an ocean: over here, the island of us; over there, the island of them. Except that a fence, unlike an ocean, can be moved. You can put it where you see fit. Likewise, if its location is unacceptable, a fence can be removed. And where are you then? How can two distinct classes exist—good, bad, desirable, undesirable—if there is nothing to delineate them?

These were some of the thoughts running through Lloyd Case's mind one morning back in 1992 when, while hunting pigs in the forests of the Pu'u O 'Umi Natural Area Reserve, above Waimea on the north side of the Big Island, he encountered a fence that had not previously been there. In recent years, state and federal land managers had begun fencing off large tracts of particularly sensitive forest to protect them from the activities of feral pigs. Whether the feral pig properly qualifies as an alien species in Hawaii is a matter of heated cultural debate. All agree, however, that pigs arrived on the islands only with the aid and settlement of humans: introduced first with the Polynesian colonists, some fifteen hundred years ago, as docile farm animals; and again in 1778, when Captain Cook released a pair of British pigs into the Hawaiian forest to run free and proliferate. The feral pigs that run wild in the forest today are the descendants of one strain or the other, or, some say, a hybrid of both.

In any event, they are abundant in number and, to many biologists, in offense. Weighing upward of two hundred pounds, an adult pig is easily capable of pushing over a tree fern and, with a formidable set of tusks and teeth, ripping out the starchy core to feed upon. This tends to kill the tree. In many cases also the rotting trunks become collection basins of stagnating rainwater, which in turn are ideal breeding grounds for introduced, disease-bearing mosquitoes. One of Carter Atkinson's upcoming projects would be to drain all the hollowed logs in a two-square-mile patch of rain forest, and then go in and trap for mosquitoes and see if this reduced the population. ("Oh God, it's gonna be . . ." He laughed as he told me. "You gotta walk back and forth, back and forth—it's really thick and uneven, no trails, you know, no roads or anything—trying to find these logs and drain them.") In addition to killing native plants and spreading foreign mosquitoes, the feral pig is believed to abet the invasion of numerous troublesome introduced plants—the fruited, strangling vines of banana poka; the thorny brambles of Himalayan raspberry—through its wide-ranging travels, digestions, and droppings. To many biologists, the feral pig is a chief culprit in the continued undoing of Hawaii's natural fabric. Resource managers at Hawai'i Volcanoes National Park spend a considerable amount each year erecting fences around select areas of rain forest and hunting, trapping, or otherwise eliminating the feral pigs within these exclusionary zones.

The fence that Lloyd Case encountered that morning had been erected by state managers, not on national parkland but on state land that Case looks on as his backyard. The fence was built of steel rails and barbed wire, at an estimated cost of twenty-five thousand dollars a mile, and it ran right through his favorite hunting grounds. This bothered him. What also bothered him, I suspect, and what caused him then and there to declare war against this and all similar fences, was that he was suddenly made uncertain as to what side of the fence—inside, outside, native, alien—it put him on.

Lloyd Case is not the sort of man one would actively choose to contend with. He is sizable, with a wide black mustache and black hair streaked with gray. One imagines that he could wrestle an alligator, if he ever saw one. In fact, Case rarely leaves the Big Island. He left the state of Hawaii only once, to visit Los Angeles, and was so disturbed by the experience he returned immediately. "I couldn't believe it. It's all screwed up, overdeveloped."

Although he may look like a man disinclined to talk, Case enjoys the opportunity to do so. He is intense, curious, genial. The first time I met him in person, after talking to him once on the telephone, he brought me a handmade T-shirt. It was yellow, and on the front was a silk-screen drawing of a bearded, bare-chested man, spear in hand, standing alongside a boar; above the design it read *MAUNA KANE*, Hawaiian for "mountain man." Case, as the saying goes, has a lot of aloha. Even the people he does contend with—state land managers, biologists from Hawai'i Volcanoes National Park—confess respect for him.

"He's really sharp. He knows the forest and native plants."

"There are some misses in his logic, but there's real communication there."

"There are people like Lloyd Case who are really articulate; if you could get 'em on your side, you could really do a lot."

One reason to have Case on your side is that he is Hawaiian. This could mean a number of things. Although the concept of aloha implies inclusiveness (*Welcome! Everyone is Hawaiian!*), in fact it disguises a subtly graded social order. The most basic Hawaiian is the resident, someone who once lived somewhere else—usually the mainland States, but also Europe, Japan, the Philippines, anywhere—and now lives in Hawaii. The common term for the palest of this suborder is *haole*. The *Hawaiian Dictionary* supplies the following definition of *haole*: "White person, American, Englishman, Caucasian: formerly, any foreigner; foreign, introduced, of foreign origin, as plants, pigs, chickens." The term is pronounced "howly" and can be used warmly or in the pejorative, depending. This rung, although the lowest, is also the widest. Slightly higher status is conferred upon those people who can rightfully call themselves locals: people born to first-generation residents but who themselves were born and raised in Hawaii. The benefits of locality are most immediately useful when speaking at community meetings or running for political office.

At the top of the social hierarchy is the native Hawaiian: any citizen of Hawaii who can trace his or her lineage back to the original Polynesian settlers. Today there are slightly more than one hundred thousand native Hawaiians, of varying dilutions, living in Hawaii, or about 9 percent of the state population. The borders of this social territory—*haole*, local, native—are largely invisible, at least to the outside eye. A woman with darker skin and wide, seemingly Pacific features might be the child or grandchild of Portuguese immigrant laborers. A man named Medeiros might be descended from King Kamehameha himself. And yet everyone knows where the borders lie. One man who has lived in Hawaii for twenty years told me he would never dare to think of himself as truly Hawaiian. He said, "The people who forget that are the ones who get in trouble." Another woman had closed her bakery in New Jersey, moved to Kauai twenty years ago, and recently married a native Hawaiian man. She had fit quietly into the community, she said, by obeying a simple principle: "They rule."

Lloyd Case is Hawaiian in the ruling sense. King Kamehameha, the first ruler to consolidate all the Hawaiian Islands under his power, reigned from the comfort of the Big Island's lush Kohala District, which encompasses Case's home and the modern environs of Waimea. Through his mother's bloodline Case counts himself a direct descendant of Kamehameha, with all the respect and ingrained knowledge that that tie engenders. For him, hunting pigs is as much a nod to tradition as to the gods of hunger. "I still use the animals in Hawaiian ceremonies," he says. "I go out and harvest them. Or I bring one home alive, until that part of the ceremony is coming up. Just like ancient Hawaiians." Or: "We view the forest as our ancestors viewed the forest. If the numbers of animals are up, we take them down, naturally. I believe there is no need to eradicate any animal; all you need is a balance."

The animal that Case sets his clock by, however, as the majority of his relatives and neighbors do and have done for several generations now, is cattle. In 1793, the British explorer Captain George Vancouver presented several head of cattle to Kamehameha as a gift; the cattle were set free on the Big Island and a *kapu* placed upon them such that none could be killed for ten years, in order that they might proliferate. And proliferate they did, multiplying so quickly that their churning hooves and insatiable appetites soon constituted a force of destructive nature. Eventually Kamehameha III commissioned John Palmer Parker—a former whaler from New England who had arrived in 1809 at the age of nineteen and jumped ship at the Big Island—to control the situation. Parker erected fences, bred a true line within them, and shot the feral animals without. Cowboys from Mexico and Spain—paniolos, a corruption of the Spanish word *Españoles*—were imported for hired help. After marrying one of Kamehameha's granddaughters, Parker acquired more than two hundred thousand acres of prime grazing land near Waimea; this became the foundation for Parker Ranch, today one of the nation's largest cattle ranches. By the turn of the century, Waimea and the ranch were the breeders of legend: paniolos, steel guitars, dusty gals. Around this time, Dr. Leonard Case moved to Honolulu and then to Waimea from Norwich, Connecticut, and worked on the ranch as a veterinarian. "He taught me that all diseases that came to this land, they were brought in," Lloyd Case says of his father. "Just like the brown tree snake." The younger Case rises early for his job in local construction. He returns home well after nightfall and falls asleep shortly thereafter.

"People say Waimea is beautiful, but it's nothing like what it was," Case said one evening over a beer in a local restaurant. "There were few cars, everyone knew everyone else. You'd tie your horse up outside. It was such a beautiful place. Then everything changed. It's happening too quick. A lot of old-timers, I feel sorry for them. They can't adapt to this. They retired from Parker Ranch on fifty dollars a month. Can you believe that? They just got washed by the side. There's more roads now, more traffic, more people coming in. Mostly people who came here from somewhere else. They came to get away from development. A lot of local Hawaiians, they're in construction, so they depend on development. I hate that part. I'm in construction too. But it's a different life now. I got electric bills; I got a mortgage. Insurance alone on five cars, that's five thousand dollars a year. I've got four homes: that's seven, eight thousand dollars in insurance. It's all for my kids, because they won't be able to afford it. More and more people are selling out because they can't afford it. More Hawaiians are moving to the mainland than ever before. The kids go to college, come back, and find no jobs here. They don't come back. This community, this town will never be the same again. The future is not us."

Among his various responsibilities, Case is the newly elected president of the Wildlife Conservation Association. The WCA consists largely of hunters and folks from the Kohala District, as well as members from elsewhere in the state. The organization hosts hunting tournaments and charity events, distributes scholarships to the children of members, and holds an annual banquet. It also provides a forum for the opponents of fences. Case said, "We can turn the heat up if we want to, but I don't use power and money to influence politicians. I think they have the message to take care. We could say 'No more fences,' and cut down all the ones out there. And you can't do nothing—it's our land! But we want to save something. These environmentalists will never know that feeling. I've got fifteen hundred years here. You've walked from one stolen land to another stolen land. You've got no history. They've got people in the hotels trying to teach aloha. You can't teach aloha! You have to live it. I always tell biologists, 'Before you fix this place, fix the place you came from; it's fucked up.' I've talked to people from New York, places like that. I give them credit. I couldn't adapt to a place like that. I go to Oahu, I can't wait to get back."

Case's main house, where he lives with his wife, three daughters, and three sons, is a pink ranch situated on a quiet street in Waimea a couple of blocks uphill from a public park. Thirty-eight years of hunting trophies are nailed around the inside of his garage: a seemingly limitless array of jawbones of boars, tusks and all. There are hundreds of them, so many that they nearly fill an adjacent room. It is the largest collection of boar jawbones in the state. Once, fifty or so were stolen; Case indicated the empty nails that formerly held them. "They were worth about a hundred dollars apiece. Since they stopped using ivory from elephants, the price has gone way up." Also in the garage was a walk-in freezer. Case opened the door to reveal three pigs, skinned to their subcutaneous fat and hanging by hooks through their lower jaws, and two enormous sides of beef.

A multicolored Playskool swing set was set up in the backyard. Several free-ranging chickens and a duck were investigating the grass around it. The odor of dog was overpowering. Farther back Case kept a series of kennels and chicken coops and small wooden houses for fighting cocks, and from everywhere came a cacophony of barks, clucks, yips, and squawks. The king of the yard was a wild boar. I had never seen one before. It was domesticated, insofar as it was castrated and lived within a hog-wire pen awaiting the social occasion that would require its slaughter. In no sense was it cute, in the way one typically pictures barnyard animals. This boar was hairy and black and very large, maybe three hundred pounds. It was an impressive grotesque, like a small rhinoceros.

As it happened, I did not see the boar, or visit the house, until after Case invited me to accompany him on a pig-hunting expedition. I did not see it, in fact, until the moment of our return. Had his invitation come a day after my introduction to the boar, I would still have said yes, but I would not have sounded so eager.

Of all the years that land managers might choose in which to erect a fence to exclude pigs from a tract of state forest, 1992—the year the Pu'u O 'Umi fence suddenly materialized, without prior public notification, in the Kohala highlands—was a particularly poor choice. The following year would mark the hundredth anniversary of the bitter fall, in 1893, of the Hawaiian monarchy; it was an eve of unusual cultural and political fervor, and blood was on everyone's mind.

The matter of the Hawaiian Islands had simmered throughout the nineteenth century. The native population, of residents born in Hawaii and linked by blood to the original Polynesian settlers, had declined by exodus and disease to a mere fraction of its pre-Cook numbers. Meanwhile, the islands' value to outsiders—American whalers, American missionaries, American managers of pineapple and sugarcane plantations—had risen steadily. The opening of Japan to trade placed Hawaii at a strategic midpoint between the Far East and newly established California ports. In Pearl Harbor, Honolulu had the makings of a central Pacific naval base. Hawaii was an independent country, feted and formally recognized by the major nations of the world. But its political base was shrinking and its military nonexistent. It was a light canoe in heavy seas. When a French warship made threatening advances in Honolulu Harbor, the monarchy began considering its options. Congressmen in Washington, fearing a British-Hawaiian alliance, debated the outer limits of manifest destiny and the ethics of declaring outright ownership. ("It is no longer a question of whether Hawai'i shall be controlled by the native Hawaiian or by some foreign people, but the question is, What foreign people shall control Hawai'i?") In 1893 a white militia, largely supported by plantation owners, ousted Queen Lydia Lili'uokalani from her palace. She was the last monarch to reign in Hawaii. The island nation was quickly annexed by the United States; in 1898 it became a U.S. territory and in 1959 a full state.

From the distance of the mainland, the historical plight of the native Hawaiian looks similar enough to that of the Native American that the two ethnic groups are often viewed as one. Their lawyers today frequently share notes. But a native Hawaiian is quick to emphasize that with all due respect, he is not natively American. Before 1893 Hawaii was marked on world maps as an independent nation. That this is no longer so is not merely unfortunate, or unseemly, or immoral, or whatever adjective one applies to the Native Americans' situation; rather, by the laws of the United States then and now, it is outright illegal. In 1893 President Grover Cleveland declared as much, decrying "the lawless occupation of Honolulu under false pretexts by United States forces." In 1993 President Bill Clinton offered a formal apology. The fact still simmers in the veins. In the past two decades, various groups have formed to support the creation of a "Native Hawaiian government." There is little question of this government's legal right to exist. However, innumerable votes and electoral conventions have yet to define precisely what shape it should take or whom it would include.

One leading faction, following the example of the Native Americans, supports a nation-within-a-nation model: Hawaii would remain a U.S. state, and native Hawaiians would form a government under the umbrella of—but not subservient to—the federal one. Other factions argue for complete sovereignty and outright secession. Peggy Ross, the leader of Na 'Ohana O Hawaii, a group based on the Waimea side of the Big Island, maintains that legally Hawaii is still a monarchy and that she herself is genealogically entitled to be its queen. In 1980 her group issued a Declaration of Independence from the United States, and she told the local papers, "This nation-within-a-nation business is just another case of giving Hawaiians half a loaf and making them pay for the whole thing." Followers of this faction refuse to pay federal taxes; they drive without license plates and have burned their Social Security cards. There is also a middle-of-the-road faction, derided equally by the groups on either side.

The stakes are enormous. With the annexation of Hawaii, two million acres of land were seized by the U.S. government and, in 1959, returned to state officials. Theoretically, all this land would again belong to the native Hawaiian government, just as soon as that entity declares itself. Much of the land has since been developed and today includes multinational quarrying and mining operations, the Pacific naval headquarters, state wastewater treatment facilities, state forests, airports, cemeteries, auto dealerships, and the Prince Kuhio shopping center in Hilo. The prospect of wealth appeals to many native Hawaiians, who have been largely passed over by the state's economic rise. (As a group, native Hawaiians suffer high unemployment and have the worst health and education profiles in the country, second only to Native Americans. On the Big Island, the disparities play out in the topography. Twin pillars of Caucasian wealth—reclusive white scientists and billion-dollar telescopes on the peak of Mauna Kea; exclusive parkland, antihunting biologists, and a multimillion-dollar tourist industry on Mauna Loa—dominate every lowland vista.) Would ownership of this real estate suddenly revert to a native Hawaiian government, or would its current owners have to pay exorbitant sums to keep it? No one yet knows, and everyone who isn't a native Hawaiian is anxious because if and when a time comes to vote on the matter, only native Hawaiians will be allowed to do so. And who is a native Hawaiian? No one can say that either, with any certainty. Various definitions are in circulation: 50 percent Hawaiian blood, 100 percent Hawaiian blood, any trace of Hawaiian blood. All anyone can agree on so far is that, since any legal definition of Hawaiianness can be made only by native Hawaiians, the final decision will be contentious.

In the natural landscape of native Hawaii, no animal is more central than the pig. It is a key figure in the Hawaiian creation myth as a source of food and ritual. Some biologists contend that the feral pigs now running through the rain forests bear no behavioral relation to the docile animals imported by the Polynesians, and argue that pig hunting is a recent cultural acquisition, no more distinctively Hawaiian than the rifle or the sport-utility vehicle. Try telling that to Lloyd Case. "The Hawaiian and the Indian are so much alike. The pig is like the buffalo to the Indian: if their numbers go down, we suffer too. The pig is chemical-free; it's better for your health than that stuff you buy. If you take it away, you tear one more page out of our history. I know there's controversy, that the pig is not the same pig as the Polynesians brought in. But look at us: part Portuguese, part Chinese. There is still Hawaiian blood in us. This pig is descended from that one."

In the early 1990s, when the Nature Conservancy of Hawaii began to trap out and eradicate feral pigs from a preserve on Molokai, hunters there denounced the effort as "eco-imperialistic." (People for the Ethical Treatment of Animals briefly stoked the resentment with a press release exclaiming, "When animals who have been on Hawaii for 1,500 years don't behave precisely as TNC would like, 'diversity of life' becomes 'alien species.'") One Hawaiian woman told public radio, "If you're going to call the pigs alien, you might as well kill us too." It was Case's daughter. Suffice to say, I had not been in Hawaii very long before I understood that the phrase *alien species,* used loosely and in the wrong company, might be hazardous to one's health.

At just past seven o'clock in the morning Case swung open a cattle gate and drove up through the dewy, hillocked pastures of Parker Ranch. He was driving a large pickup truck and pulling a red, wooden-slat trailer that held his hunting dogs: Roxie, a black female Doberman; and Spotty, a mix-breed male with black and white spots. Case wore jeans, sneakers, and a T-shirt with the sleeves ripped off. The absence of sleeves revealed thick cordons of muscle and a large and intricate tattoo on each shoulder blade, octopus on the left, boar on the right.

Though dawn was barely thirty minutes old, the cows were busy ruminating. Case drove on, past a reservoir that tapped into various streams running from the mountains we were headed into, until he reached a padlocked gate. He had forgotten the key, so we parked there, released the dogs, walked around the gate, and began hiking up a dirt road into the forest. For a knapsack Case wore a burlap coffee sack tied tight with rope; the shoulder straps were lengths of rope, covered, for comfort, with a layer of foam and old carpeting. Presently the sound of an engine approached from behind, and soon a pickup truck pulled up alongside. It was Case's friend Harry Wishard, with his wife and daughter along for the ride.

Wishard asked, "Why aren't you driving?"

Case smiled. "I like to walk. The farther the better."

Wishard waved and drove on; Case kept walking.

It was two hours by foot uphill to prime hunting grounds. Case passed the time amiably expounding on pigs, Hawaiian life, and the boundary between natural and unnatural. He bent down to examine tracks in the mud. He cupped his hands and drank from a small stream. At times I felt like I was in an old Injun film; perhaps Case did too.

"What people don't understand is the pig is a sacred piece of our culture.

"One of the things the scientists wanted to do was kill all the earthworms because earthworms are not native to this island. But what earthworms do, their basic function, is to turn organic matter back into nitrogen. Whether it is native or not native, don't get rid of 'em.

"When the ocean is rough, the pig is fat. When the ocean is smooth, the fish is fat. Hawaiians know—the seasons never change.

"I try to go hunting once a month, but I don't like to go out when the sows are giving birth or pregnant. The modern hunter goes anytime. He'll take eight or ten dogs. It's ridiculous. They'll go out and kill sows and piglets—that's killing the future. One hunter came over from Oahu with fifteen dogs! Some people go for trophies. I don't call them hunters."

There are several hunting groups in Hawaii aside from Case's, many members of which are not native Hawaiian. What all the hunters share is a dismay at the gradual expansion of the national parks and state reserves in Hawaii over the years, and the gradual dwindling of available hunting grounds. One Sunday morning I attended a meeting of Pig Hunters of Hawaii, which was founded in 1992 after the Pu'u O 'Umi fence went up. The meeting was held in the back room of the McDonald's on the main thoroughfare in Hilo, about a mile from the airport. The restaurant was located directly under the flight path and shook from time to time with the roar of departing jets. The attending members included several hunters from the Hilo side of the island, as well as a fellow with long black hair and mirrored sunglasses from the Kohala region, where Case lives and hunts. In addition, they had invited two members of the Hawaiian Rifle Association, based on Oahu, and a California representative of the National Rifle Association, on whom the group hoped to impress its grievances.

"All this land has been removed from public use, or is scheduled to be removed. They say it's for the public good, but the public isn't invited in; it's for their own use. Like Hawai'i Volcanoes National Park: they tell you it's a great success story, but they still got weeds in there."

"When you come into the visitors' center, there's a stuffed pig—all this about how they're destroying the forest, the worst thing that ever happened to the park. And they go around telling our schoolchildren this. The propaganda is incredible."

"I do want to save native species and native ecosystems. But we're doomed as far as hunting goes."

"We feel that all animals that are here should be protected. Not eradicated, but kept in balance. That's the thing. There are a certain number of animals here that can coexist with the native plants and animals."

"My impression of the Endangered Species Act is it's a bunch of bullshit. They've got a little green plant on the beach on Molokai? It's an endangered species, but I got it growing all over the place at my house. I'm whackin' it up the wingwang with my Weedwacker!"

"Hawaii is losing its identity," Tom Lodge, a longtime member and a sightseeing tour pilot, said afterward. "The plantations are all gone. The rural lifestyle is being encroached upon. The fears of the environmental community about the forest are very real: alien plants, bugs, development. The threat is very real, and we all want to do something, but they don't feel that the hunters are credible. Our view is that some of these problems are being in kept in check by the animals here." He added, "If you start saying we should have English only, the response is, 'Oh, no, we'd be taking away from people's culture.' Yet if you ask them about the introduction of a plant or animal? 'Absolutely not! Why would you want to pollute the environment?' Like with the brown tree snake. What if it swam here? Do you deny its right to existence if it got here by itself? If some bird species got blown here? What's here is here. It has a right to be here. If life showed up from outer space, would you preserve it? Or would you go out there and Clorox it?"

When Case and I at last reached the Pu'u O 'Umi fence line, there was little to see. Shortly after the Hawaii state legislature learned of the construction of the fence, as local hunters made certain, state land managers tore it down. Now all that remained was a two-foot-wide corridor of air through dense bracken and moss-draped hapu'u ferns. Every hundred yards or so sat a heap of rusting galvanized steel rails—former fence posts—that Case complained were now leaching zinc into the groundwater.

For a while, the state mandated a series of meetings between the Hawaii Department of Fish and Wildlife and the Kohala community, the tenor of which, early on, was heated to say the least. From these meetings another council was born: the Natural Areas Working Group, made up of folks from the community, the national park, and various state and federal agencies with the collective aim of reaching some consensus on which areas of the Big Island could be fenced to exclude pigs and which could not. In addition, the hunters had a few questions of their own: What exactly distinguishes a native forest? What makes a forest "pristine"—and is that according to pre-Hawaiian standards or pre-European ones? And most important: Why are outsiders making decisions about what is or is not alien to Hawaii?

The life span of the working group lasted through several dozen meetings. Case attended when he could, but it is his firm contention—and the contention of virtually every hunter with a view on the matter—that the fencing out and removal of pigs from forests has not slowed the influx of alien weeds. If anything, he said, it has exacerbated the problem. "I use the knowledge from when I was young. And now I can see the difference. What scientists say—the forest has gotten better? No, it hasn't. I see now, with the reduction in the animals, it has gotten worse."

Having located the old Pu'u O 'Umi fence line, we now began to follow its ghost of a trail farther into the Kohala highlands. The sun had disappeared behind a veil of drizzle. The bracken was sopping. The mud, at times, sucked to the knee. From a hillside, Case pointed across to Waipio Canyon, a green cliff line that dropped precipitously to the seacoast below. A lush plateau at its top was the heart of the pigs' breeding area, Case said. He pointed to a series of barely discernible wavering trails cut by pigs migrating up and down the canyon wall. The plateau was also the site of the state's next embattled proposal, a plan to fence one hundred twenty acres of a larger area called the Bog Unit. Stiff offshore winds sweep up the valley walls, annually depositing one hundred fifty inches of rain on the plateau, thoroughly saturating the soil. The Bog Unit encompasses two adjacent montane bogs, each with a distinct and rare assemblage of plants: native sedges, grasses and mosses, an endemic violet, as well as various native trees so stunted by the conditions that in some cases they grow no more than a foot tall. The unit is also home to several endemic birds, a few weeds, as well as mongooses, rats, feral cats, and migrating pigs. Not incidentally, it would sit squarely on the pigs' migration path. Although pig hunters rarely hunt in the bogs—a rough three-hour hike is required just to reach it—they would consider any fence there as a veiled attempt to cut off their supply.

A draft environmental assessment of the fencing project notes that "feral pigs are present in high numbers in and around the project area. Their activity has contributed to the destruction of native vegetation and subsequent invasion by nonnative weeds in much of the area surrounding the bog. Large areas have been converted from native shrubland into meadows of alien pasture grasses . . ." Case, slogging on into the thickening drizzle, said simply, "A fence don't belong in the forest. I don't care what anybody tells me. It ruins the natural beauty." Later he added, "If fencing were such a great idea, the world would be covered with fences."

10

Consider an experiment set in a remote rain forest on an island in the ocean. In outward appearance, this experiment consists of no more than a fence—a mile or two of barbed wire stretched from steel post to steel post, extending beyond sight in both directions and eventually turning on itself to enclose several hundred acres of forest. It is not so much a fence erected for conservation as much as a fence erected to test the validity of conservation. As a boundary or border, the fence is essentially porous: forest birds can easily pass from outside to inside and back again; also insects, flying or crawling. Even snakes could come and go, although to the best anyone can detect, there are yet no snakes living in the forests of Hawaii. The same plants that grow inside the fenced area also grow outside it: the same trees, tree ferns, epiphytes, and ground-dwelling lilies; the same mosses and fungi; the same dead leaves and forest litter; the same soil and muck; the same worms, soil mites, and microscopic muck dwellers. In fact, the barbed-wire fence is impassable to only a single species, *Sus scrofa*, the feral pig. The fence has been erected specifically to prevent any feral pigs from crossing into the enclosed area; in that sense, the enclosed area is less an enclosure than an exclosure, an exclusionary zone. Whether the forest inside the exclosure, where no feral pigs roam, is measurably different from the unprotected forest outside the exclosure—whether, say, its trees produce more or fewer leaves, or its soil is more or less fertile, or the worms, flies, mites, and other soil dwellers are more or less abundant—is what David Foote, one of the principal scientists who designed the experiment and erected the fence, aims to find out.

Foote is a biologist at Hawai'i Volcanoes National Park; by training, he is an entomologist, a studier of insects. He stands at medium height, with boyish blue eyes and sandy, inherently tousled hair. His voice, warm and precise, bespeaks a loving devotion to the millimetric. I would have left that impression to characterize him, until I caught sight of him early one morning in his backyard. The mist had not yet lifted, and Foote, in knee-high rubber boots, was strolling around in it, Patton-like, with a cigar clamped between his teeth, appearing less the investigator of multilegged microscopia, more a strategist of large-scale tactical combat.

Foote lives on the Big Island in the village of Volcano, a small grid of quiet residential roads two miles down the road from the entrance to Hawai'i Volcanoes National Park. The more remarkable features of Volcano include a diner, a gas station, a new hardware store, and a persistent drizzle. Sitting at an elevation that corresponds exactly with the cloud line, the village receives one hundred forty inches of moisture a year, forty inches more than the park immediately uphill. Moss grows on Foote's welcome mat. A row of tall trees obscures his neighbors; a gravel road runs past. Foote lives there with his wife, Karin, and their young son, Sam. Also in residence are two jittery dogs: Homer, so named for his odyssey through three prior owners, and Sed—an acronym, Foote explained, for Someone Else's Dog.

Although insects occupy a great deal of his laboratory and field time, Foote, like most biologists who work in the Hawaiian rain forests, is increasingly preoccupied with the movement and impact of feral pigs. A colleague of Foote's who grew up in Hawaii and has worked at the park nearly three decades, explained it like this: "Once you get pigs in a forest, they cut up the roots, trample the soil, and the soil loses its air space. It loses the natural aerobic processes. The production of the roots of trees diminishes. The productivity of the forest is diminished; reproduction is diminished. The structural diversity is greatly diminished. The structural elements for retaining moisture are gone. When you lose your groundwater, everything else gets affected. The plants that are vulnerable—which are many—are diminishing in abundance. If you look at the endangered species list, all the plants are either edible or highly fragile. Even the trees have edible seeds."

However, for all the effort spent on eradicating and excluding pigs from sensitive habitats, the animal's true ecological impact has gone largely unmeasured, its notoriety more a product of suspicion and estimation than of precise measurement and analysis. "People haven't done a good job of quantifying it," Foote says. "Botanists tend to believe that pigs eat every rare plant in the forest, yet you wouldn't be able to find a single study in the literature to demonstrate that. It's not so much that I don't think that pigs cause damage, so much as I think that conservationist biologists have done a very poor job of demonstrating their case."

Foote means to confirm or deny the validity of those suspicions. With Peter Vitousek, an ecologist at Stanford University, Foote has established four pig-free areas of Hawaiian rain forest: two on the Big Island, within a few miles of Foote's office, and two on the neighboring island of Molokai. He collects samples of soil from these areas and returns them to his laboratory for analysis. The lab sits directly across the hall from his office. It is a white, fluorescent-lit room with a counter, microscopes, and a row of metal stools along one wall; along another wall, a bank of windows looks out onto the cluster of drab wooden buildings that comprises the park headquarters. When I stopped in, Christmas lights—the oversize kind, in blue, yellow, green, red, and white—had been strung across the ceiling in parallel rows. We were still only in November, a bit premature, it seemed to me, for yuletide festivities.

"Dime-store biology," Foote explained. Rather than pay fifty dollars for a device known in scientific catalogs as a Berlese-Tullgren funnel, he had engineered his own, in quantity: plastic jugs, their bottoms cut off, that would hang upside down directly under each bulb; a glass vial would hang under each spout, secured there by a rubber band and the cutoff neck of a balloon. Already an intern was busy hanging the first of one hundred sixty jugs. Each one contained a sample of rain forest soil from a field site on Molokai. In three weeks, heat from the bulbs would dry out the soil, and the soil's inhabitants—the centipedes and millipedes, mites, springtails, pill bugs, isopods, arthropods, and other, smaller invertebrates that Foote identifies collectively as "micropods"—would migrate down to the spout and into the vial for later study. As a group, the micropods would form a measure of the productivity of the soil and, by extension, of the rain forest immediately surrounding the area from which the soil sample had been taken. These measures in turn would go some way toward revealing whether a forest without pigs is quantifiably better off—more productive or functionally efficient—than a forest overrun. Looking up at the downward-pointing spouts, trying to picture the descending micropods, I understood that my quest for unfiltered nature was leaving familiar, macroscopic terrain and entering a far finer scale. Where gravity led, we would follow.

Usually it is the job of interns and assistants to go over to Molokai for a couple of days, traipse around in the rain and muck, and come home with the requisite bags of dirt and bugs. Foote likes to keep an eye on things, though, so one evening he boarded a small commuter plane at the Hilo airport and flew over there himself—to have a look, collect a few insects, and all in all continue his efforts to understand how the ecology of Molokai relates to that on the Big Island and on Kauai, where he has also done some fieldwork. Insofar as all the islands in the Hawaiian archipelago stem originally from the same volcanic hot spot, they represent the same geological and evolutionary processes seen at vastly different stages in time and development, and they invite comparison. As the plane landed at the darkened airstrip on Molokai, Foote said, "So much of the focus is on the unique—that's the national park strategy, and it works well. But what's interesting to me is the interconnectedness of environments. I like the idea that you can come to Hawaii and see the same basalt lava that's outside your home in Los Angeles."

A battered four-wheel-drive truck awaited us in the airport's small parking lot. Foote scrounged around inside for the keys, then drove out of town. After a few miles he turned onto a narrow dirt road leading up into the foothills of a cluster of low, formerly volcanic mountains that occupy the eastern half of Molokai. In the morning, we would drive to the top, to Kamakou Preserve, several square miles of pristine forest in which Foote's experiment was somewhere nestled. For now, our evening stopover was an abandoned Boy Scout forestry camp, a few miles shy of Kamakou proper, that serves as base camp whenever Foote or his interns pay a visit to the field site. It was well after dark when we pulled in. Through drizzle, our headlights illuminated a patchy lawn. Nearby was a small shack, from which soon emerged the property's caretaker, a shirtless fellow with a long gray beard and a large, hairy belly: Santa at the bleary end of his career. He greeted Foote warmly and handed him a set of keys. Foote crossed through the dark to a ramshackle wooden barracks; he found the fuse box, unlocked the front door, and ventured inside.

When I entered a few moments later, Foote was busy in the kitchen, such as it was: a bare bulb illuminating a picnic table, a yellowed refrigerator, curling linoleum floor tiles, and a porcelain sink crawling with cockroaches of astonishing size. The place seemed to manifest a kind of limbo, an intermediary realm between the fully modern world of sidewalks and street signs and the classically natural world of forests and, at most, footpaths. Indeed, I was inclined to think of the cabin as a kind of frontier, a stepping-off point—forgetting that mine was the human perspective and that in fact the cabin was as much an entry point as an exit, a haven for the various animals that, like humans, prefer not to have nature too close around them. Foote wandered from the kitchen into the sleeping quarters, a large, open room with a dozen or so steel bed frames and old mattresses. He found a bed to his liking and tossed his sleeping bag onto it. He advised me to pick one not located directly under any of the rafters, which he referred to collectively as "the rat highway." I fell asleep to the whine of bloodthirsty mosquitoes and the patter of rain on the tin roof.

Morning brought clearing and a slight breeze. After a quick breakfast—hard-boiled eggs, toast, and instant coffee—Foote loaded gear into the truck and began the drive up to Kamakou. The road wound and rose through a forest of towering eucalyptus trees, imported and planted there decades ago to replace the massive stands of native sandalwoods that had been cut in the early nineteenth century to feed a short-lived sandalwood trade with China. Back then, perhaps, the road had been easier to navigate. However, subsequent generations of mechanical smoothers and graders had succeeded only in scraping the road several feet into the earth, producing something akin to a clay slalom. From one rise, Foote surveyed the stretch ahead: a steep downward slope that cut left, then sharply right again at the bottom just before passing over a narrow, unrailed bridge—from here to there one long mudslick. He gripped the wheel. "It's kinda hairy," he said, and smiled tightly.

Foote's initial entrance into entomology involved a similarly controlled slide. His first real taste of field biology came in the summer after high school, when he took part in a study of a herring gull colony along the Massachusetts coast. For two weeks he walked the shore with binoculars, reading the leg bands on gulls and conducting a population census. "The idea of being able to follow a cohort through time was very intriguing to me," he said. The following summer, Foote focused his bird-watching skills on a gull colony on Thrumcap Island, in coastal Maine, observing the different postures and calls: what, when, how often, why. "It was my first experience in actually quantifying anything in biology, and convincing other people that what you're seeing is indeed a pattern." Foote was nineteen at the time. In college, he gravitated toward the zoology department at Berkeley. When it came time for graduate school, he sought out a traditional ecology program and found it at the University of California at Davis. One adviser was Jim Carey, a leading researcher in California's long-standing effort to protect its agriculture from successive incursions of foreign fruit flies. "I didn't know anything about entomology at the time. That was my first chance to do serious population demography—and that's when I began to realize what some people go through to get tenure. It was very discouraging to watch him just grind out these life tables one after another."

Life tables are to biologists what human mortality schedules are to insurance agents: mind-numbing assemblages of life-expectancy data. Place a group of insects in a cage, all of a given age. How many die each day, how many new ones are born, how quickly does the population grow or dwindle under various conditions? In California, where introduced fruit flies are a nagging problem, fruit-fly life tables are a valuable commodity. The state agriculture department has devised a clever strategy to check the spread of pest flies: breed sterile flies in the lab and release them en masse; when the wild pest flies mate with the sterile ones, they produce no offspring. For the plan to work, state biologists must be able to pump out several million sterile pupae in the space of a week. This requires a detailed understanding of the fecundity and mortality rates of the fly in question—which in turn requires an entomologist somewhere, or his research assistant, to devote a significant proportion of his career to watching caged flies live, breed, and expire. "It's not that it's not meaningful information," Foote continued. "It depends on the sorts of questions you're asking. What bothered me was that it's so divorced from the reality of the organism's true life cycle. Patterns of mortality in the natural environment can be completely different than in a controlled laboratory. I could see how doing one life table . . ." He broke off to consider a fork in the road.

"I think this is the right turn, watch me if I'm wrong." A dozen yards ahead a road sign appeared: NO TRESPASSING. Foote laughed. "No, I think I'm wrong. Boy, and I thought I knew my way."

By now the road had risen through the area of planted eucalyptus and had entered native forest: stately, slender koas and o'hias, the hallmarks of a woodland on the drier side of a Hawaiian island. The landscape changed again in a few moments as the road climbed a little farther. Foote parked the truck in a cul-de-sac. The air was cool, the foliage bright with moisture. Foote grabbed his knapsack from the back of the truck, then led the way down a narrow trail through a forest of bracken and the mossy trunks of budding tree ferns. After a few minutes the path opened out onto a sunny, open plateau. A stiff breeze sent clumps of mist tumbling at us. The ground was sodden, the vegetation low and bristling: knee-high ferns; fountains of long, narrow, spiked leaves belonging to an endemic lily of the genus *Astelia*. In short, a montane bog, stunted by acidic soil and constant exposure to a soggy breeze, the whole of it creating the appearance of a terrestrial seabed, a field of anemones and flowering barnacles and green, many-tendriled corals combing the tide of moisture that ebbed and flowed across them. Between bouts of mist it was possible to see the ridge opposite, and below that, sheer cliffs of foliage that dropped two thousand feet to a merciless coastline.

In 1986 Foote's work with fruit flies brought him under the wing of the California Department of Fish and Game. He had given little thought to state of Hawaii, or to the state of Hawaiian entomology, until one morning when he drove to the marina in Sausalito to pick up the department's single outboard. Moored next to him, he remembers, was a gaff-rigged sailboat owned by a young couple. "They were fixing it up and fixing it up. One day we got to talking, and it turned out they were planning to sail to Hawaii. Seemed like a long way to me, but they said nah, it's easy. One day they were gone. I don't know if they ever made it, but that got the bug in me."

The bugs experienced by Foote are now beyond counting. More so even than the birds and snails, the most prominent and evolutionarily radiant members of Hawaii's land fauna are the insects: there are at least ten thousand native species, half of them yet unnamed, with untold more yet undiscovered. Thirteen hundred species of beetle. Ninety-three species of mites and ticks. Seventy-five species of bark lice. A hundred and fifty-four species of spider, nearly all of them of the ballooning variety. There are sapsuckers and leaf miners, wood borers and lichen eaters, decomposers, hunters, kleptoparasites. There are flightless lacewings, flightless moths, flightless crickets, flightless wasps—even, entomology's oxymorons, flightless flies. There is a weevil that grows algae on its back as a disguise; there is a wasp that walks under rocks in swiftly moving streams, where it lays its eggs in the pupae of beach flies. All of them are descended from no more than four hundred ancestral species that colonized the islands sometime during the past seventy million years—four hundred winners of "the dispersal sweepstakes," as biologists refer to it. Just as notable are the losers, the species that did not make it. There are nine hundred fifty native species of moth in Hawaii, but only two butterflies. A hundred and five native species of housefly, but no horseflies or deerflies. Six hundred fifty-two native hymenoptera, including sixty-four species of yellow-faced bees and thirty-three species of square-headed wasps, but no bumblebees. No native cockroaches, no termites, no ants. The only native species of flea, *Parapsyllus laysanensis,* clings for its life to seabirds in the Northwest Hawaiian Islands.

But all that is changing. There are now some twenty-seven hundred alien species of insect in Hawaii, including six hundred forty species of beetles; six hundred twenty-four ants, bees, and predatory wasps; twenty cockroaches; and nine fleas. Some arrived inadvertently; others were intentionally released—decades ago by the Hawaiian Sugar Planters Association, and later by the Hawaii Department of Agriculture—to combat the inadvertent ones. Insects arriving, insects going extinct, insects who knows where or what to call them. Foote says, "You fog a tree with insecticide for five minutes, and dozens of species fall out." It could be argued that ornithologists have the tougher emotional task, that whatever anthropomorphic calculus ranks birds more sympathetic than insects—larger eyes, perhaps, widening onto minds more easily considered—also binds the prospect of their extinction that much more tightly to the human heart. The entomologist, however, is confronted with the weight of numbers. There are more insects than birds: more of them present, more of them passing. Foote knows that when the time comes for him to leave Hawaii, there will be fewer native insects than when he arrived. He also has seen the toll that extinction exacts from biologists too long exposed to it. So Foote has armed himself. With transects and dendrometer bands, pan traps, gas chromatographs, ion-exchange resin bags, with a practiced detachment and a slingshot, he awaits the green unraveling. "I take the position that it's going to happen, and I'm going to document it."

Foote continued across the fog-swept bog until the trail reentered the forest, an open woodland dominated by native o'hia trees. The trail went a mile or so farther in, toward Foote's eventual destination, the fenced pig-free zone that was the source of the soil hanging in plastic jugs beneath

Christmas bulbs back in his lab. For the moment, though, he busied himself with catching flies, which provide yet another gauge of the changes under way in Hawaii's forests. Sun filtered down through the trees, rustling in a steady wind: wonderful weather for humans, Foote said, but terrible for flies. "These are the worst possible conditions—I just want to warn you." Of all the Hawaiian insects, the dearest to Foote is the family Drosophilidae, the pomace flies, better known, not entirely correctly, as fruit flies. Included in the family Drosophilidae is the genus *Drosophila*. Among genetics researchers, *Drosophila* is the organism of choice. Its chromosomes are few and big and easily extracted from its salivary glands; it reproduces at ten days of age, so genetic changes unfold observably from generation to generation, which means you can wrap up an experiment in a matter of weeks. Also, flies don't eat much—not like mice or lab rats, which will quickly chew a hole in one's departmental budget. Over the past century, the science of genetics has grown up around one drosophila species in particular, *Drosophila melanogaster*, a tiny red-eyed orange fly so ubiquitous that scientists simply shrug and call it "cosmopolitan." *Drosophila melanogaster* is a guinea pig with wings. It has been scrutinized and forcibly mutated, crossbred, back-bred, inbred. Scientists have created drosophilas with extra-long life spans (sixty days, instead of thirty), drosophilas with superior maze-navigating abilities, drosophilas dumb as posts; drosophilas with no legs, with legs sprouting from their heads, even—I saw their photograph a few years ago on the front page of *The New York Times*—with extra, ectopic eyes peering out from where their knees should be. A great deal of what we know about ourselves has been gleaned from monkeying with this pale orange fly.

In contrast, the drosophilas of Hawaii owe their oddity entirely to the whims of natural selection. They are a tribe unto themselves: oversize, with elaborate stripes and colorations and strange and intricate mating rituals. To prove their sexual worth, males of the species *Drosophila heteroneura* butt heads, which are elongated like those of hammerhead sharks. Males of the closely related *Drosophila silvestris* stand on their hind legs and grapple like boxers in the clinch. If you want to understand the genetics of colonization, isolation, and speciation in a nonlaboratory setting, the Hawaiian drosophilas are your subject; they are honeycreepers for entomologists. There are some six hundred species of *Drosophila* in Hawaii, one-fifth of all the *Drosophila* species in the world—the progeny of flies that tumbled from earlier Hawaiian islands and have been doing so for forty-two million years, ever since the one fly from which they are all descended blew in from somewhere else, to an island that long ago submerged. From one strange and alluring species to the next, drosophila are themselves a sort of archipelago of biodiversity.

And as surely as the honeycreepers are indicators of environmental change in Hawaii, so too are the drosophilas, perhaps more so. Ecologists sometimes describe an ecosystem as a sort of pyramidal hotel of energy consumers, built up of successive trophic layers of feeders and fed-upons: plants, which draw their energy from light; grazers, which draw their energy from plants and span a range of organisms from leaf-mining insects to fruit-eating bats to Jersey cows; and predators, like tigers, feral cats, *Tyrannosaurus rex*, and bird-eating brown tree snakes. It is a loose schema, with numerous exceptions and outstanding questions. (Which story, for example, do carnivorous army ants inhabit in the Amazonian rain forest pyramid?) The drosophilas occupy a janitorial closet near the base of this building. They subsist largely on decaying plant material—bark, branches, leaf litter. They are composters, thriving on the senescence and misfortune of their fellow hotel guests. Most everything ends with them. The drosophilas are so ubiquitous in Hawaiian rain forests and their microhabitats so varied that their seemingly minor fates in fact closely reflect the spectrum of disruptions and alterations unfolding above and around them—including but not limited to the damage caused, or said to be caused, by feral pigs. If your quarry is the pig, it pays to follow the flies.

"The drosophilas are decomposers in this ecosystem," Foote explained to me one afternoon on the Big Island. "They're responsible for breaking down the organic matter in plants and allowing the nutrients to be cycled up into the forest again. These particular species breed on plants that are some of the most sensitive to disturbance by pigs, cattle, and rats. So we're very concerned about their status."

Foote had taken me to into the Ola'a Tract, a modest and representative stretch of rain forest just a couple of miles from Foote's office at Hawai'i Volcanoes park headquarters. A two-hundred-plus-hectare portion of the Ola'a Tract had been fenced off to exclude feral pigs. Along with the two pig-free plots on Molokai and another on the Big Island, this tract comprised one-quarter of Foote's larger investigation into the ecological impact of the pig. The foliage was dense and dripping, and Foote led cautiously along an invisible path, stepping over moss-draped logs, two thousand years of organic decay slurping at his boots. In 1963, when some biologists from the University of Texas began a long-term study of Hawaii's drosophilas—the Hawaiian Drosophila Project, as it is known today—the Ola'a Tract is where they came. Foote began working at Hawai'i Volcanoes National Park in 1990, and he has spent a considerable portion of that time tramping through Ola'a, chasing drosophilas, both as objects of interest unto themselves and as heralds of a more widespread flux.

To study or collect wild drosophilas, a researcher must first attract them; there is a time-honored method for doing so, at which Foote is well practiced. First you'll need some mushroom spray: slice up some mushrooms, let them ferment in yeast for a week, then pour the resulting juice into a handheld sprayer. (A common houseplant mister serves just fine.) It is wise to first strain the juice through a fruit sieve to remove the larger mushroom chunks that might clog your sprayer. Then, on the night before you are to begin your fieldwork, mix up some banana bait. Slice several bananas into a large bowl, add yeast, then mash and wait; by the next morning the concoction should be spilling out of the bowl onto the counter like a post-nuclear tapioca. Put some of this unctuous ooze in a Tupperware container; grab your sprayer, a butter knife, and some kitchen sponges; and off you go. Once you are in the forest, find some spots to tack up your sponges: tree trunks are good, at eye level, out of the wind. Spread on some banana bait, give a few squirts of mushroom juice, and watch what flocks in. Foote now led us to an orange sponge that he had slathered with goop and tacked to a tree trunk a few moments earlier. The odor resembled something between mildewed sweat socks and stale beer. "You get used to it, to the point where you really enjoy it in the morning," Foote insisted, with a note of apology. "You *do*. It's a very mulchy smell, like rotting leaves. The flies come to it like a magnet."

Already a flurry of drosophilas had settled. Several were of the alien variety, the small orange foci of millions of hours of laboratory tedium and common pests of the kitchen fruit bowl. The other drosophilas on the sponge, the Hawaiian ones, were darker and much larger. Foote pointed to one with a splotchy pattern on its otherwise transparent wings. Among the Hawaiian drosophilas, there exists a subset of species known as picturewings—this specimen was one—of particular interest to Foote. There are more than a hundred species of picturewing drosophila, each one endemic usually to just a single Hawaiian island. They all evolved within the past six million years, on the islands now present, from a single species very much like one still found on Kauai and named, in honor of its seniority, *Drosophila primaeva.* Through a close reading of the map on the wings of the picturewings—in this case, "the T-bar infuscation on the M cross-vein"—individual species can be identified as precisely as snowflakes. Foote said, "A tremendous amount of sexual selection has gone on in the evolution of these flies. From that pattern, I could tell you not only what species it is, but what sex it is. This is *Drosophila setosimentum.* It's one of four species in the park that breeds in the rotting bark of this *Clermontia* here."

He indicated a low sapling growing nearby; a cluster of spiny leaves sprouted from its top end, like a knee-high palm tree. The picturewings, Foote explained, lay their eggs almost exclusively in the rotting bark of this and other, often rare, Hawaiian plants, including the olapa tree and the ieie vine. One picturewing species is particularly choosy. It lays its eggs only in the rotting bark of the native soapberry, a tall, smooth-skinned tree long prized for its hardwood. Moreover, the bark must be attached to very large limbs or fallen trunks of the tree. (On smaller limbs, where the ratio of surface area to volume is high, the bark dries out instead of rotting.) And the fallen branch or trunk must initially be green—that is, it must have been broken off by the wind and then begun to rot. At the dissolving border of green bark and new rot, that is where *Drosophila engyochracea* leaves its larvae. "And it's been shown that they're feeding on bacteria and fungi that are associated with the decomposition, at the same time that they're feeding on the decaying bark itself," Foote said.

As goes the rot, so go the rot eaters. Surveys of the drosophila population have found a steady decline in the number of picturewing species over the past three decades. Four species of picturewings that were present in the Ola'a Tract in the 1970s and 1980s, including two of the most magnificent species, are now missing from the forest. Twenty years ago, *Drosophila setosimentum,* the fly that sat on the goopy sponge immediately in view, had once been among the most common species in the area. Now it is one of the rarest. Their plight, Foote said, is probably linked to the feeding habits of feral pigs and introduced slugs and rats, which have caused a general decline in the abundance of the native plants that the drosophilas breed on and eat. In 1992, he and Hampton Carson, a colleague at the University of Hawaii, came to Ola'a and surveyed the relative abundance of drosophilas both inside and outside the fenced area, as a proxy for the relative abundance of the native flora. They tacked up sponges inside and outside the fence line—some eight hundred sponges slathered with banana bait and sprayed with fermented mushroom juice. Flies came; Foote and Carson saw; they counted. Their survey revealed that the native drosophilas were denser in number inside the protected area, presumably because their host plants are more numerous there. In contrast, alien drosophilas occurred in greater numbers outside the protected area. "The alien species are much broader in their diet," Foote explained. "They're able to feed on a variety of native plants, but they're also able to feed on a large number of the alien plants that are invading this ecosystem"—plants thought to be spread, he added, by the feral pigs outside the fence. "Keeping pigs out, and keeping pigs from spreading weeds—those are the two factors at work here."

On Molokai, Foote began tacking sponges to trees along the trail and smearing them with drosophila bait. He had planned to leave the baited sponges up overnight and return the following morning to survey the attractees, but the possibility of another night of rain forced a reconsideration. Instead, he would inspect the sponges in a few hours, on the hike out, after we had visited the pig experiment. "We won't get as much that way, but the idea of coming back here at six o'clock in the morning . . . The road is marginal right now. I don't think I'd wanna be stuck here if it rains any harder." He wandered ahead on the trail with his Tupperware and sprayer, tacking up a sponge every dozen yards or so—sponge, glob of banana, essence of mushroom—and leaving each to waft its feculent aroma. "Calling all flies," he cried out.

Whether or not nature is best viewed as a pyramid, wherein the snakes, tigers, and other predators reign from the apex while the grazers and grubs wallow at the base, the important point is this: there are many more of the latter than the former. The mulch-loving drosophilas may occupy a janitorial closet in the larger scheme, but as Foote has come to learn, they are far from alone down there. There are entire sub-basements of multi-legged and near-microscopic creatures that toil ceaselessly and unnoticed, not unlike a custodial staff, breaking down the day's fatalities and recycling their nutrients into a stratum—soil—on which tomorrow's life builds itself up anew. Walking along the trail again, Foote bent down and drew my attention to a native lily growing alongside the path. Its leaves were long and narrow and fountained outward from a central stalk. Foote took a leaf in hand and gently peeled it back ever so slightly to reveal the axil, a slim pocket between the base of the leaf and the stem itself, where a small well of rainwater and detritus had formed. "Bingo!" he exclaimed.

Floating amid the leaf gunk were two tadpolelike creatures—two larvae, Foote said, of the genus *Megalagrion,* which encompasses Hawaii's damselflies. Close kin to dragonflies, damselflies are typically smaller and more slender. Moreover, whereas a dragonfly alights with its four membranous wings splayed horizontally, helicopter-like, a damselfly rests with its wings folded vertically above its back, like a fighter plane parked on the deck of an aircraft carrier. When an iridescent red damselfly paused on a nearby leaf, Foote could reach out and, by gently holding its wings together between his thumb and forefinger, lift it from its perch. As with the drosophila, the first *Megalagrion* colonist reached ur-Hawaii millions of years ago and began radiating, giving rise to a host of species endemic to one or two of the islands visible today. So far, thirty species have been identified, though how many there are in all is hard to say: until Foote came along, nobody had bothered trying to tell them apart.

"All *Megalagrion* have been lumped into a single species from the islands, which is ridiculous," he said. "Their morphologies are completely different." By collecting specimens from different islands and analyzing their DNA, it should eventually be possible to reconstruct a phylogenetic history of the genus, much like what's been done with *Drosophila:* which species are more closely related to one another; which are older; which one, perhaps, is the primeval ancestor from which the others evolved. An entomologist could collect damselflies, or try, by sweeping through the forest with a net on a pole. It is easier, however, to catch them in their youth. Adult damselflies lay their eggs in the pools of water that form in the leaf axils of plants like *Astelia,* the native lily, and the eggs soon hatch into wriggling and voracious little naiads like the ones Foote was now admiring. "They're predators in the soil and litter system," he said. "The species here has specialized to the point where it feeds only on species that crawl into a leaf axil of *Astelia.* The *Astelia* act as nets, they catch litter, and that litter decomposes and builds up a little microhabitat of organic matter in the bases of the leaves. That's colonized by springtails and mites and other soil organisms, and they in turn are preyed upon by the *Megalagrion* larvae."

What Foote was describing—what he has spent and will spend years more delineating—was an underworld ecosystem. Detrivores, microbivores, leaf-litter eaters, and leaf-litter-eater eaters, small things consuming smaller things and then themselves consumed, all the way to the bottom. Composition, decomposition: what's old is new again. When looking out on the natural world, there is an understandable tendency to focus on the plants and animals that one can most readily see—the flashy and unusual birds, the sinister bird-eating snakes, and other "megafauna" near the apex of the pyramidal food chain. As a consequence, it becomes all too easy to mistake their lofty perspective as the only overview and to project onto the landscape a sort of upward inevitability, as if all the other plants and organisms exist only, like piled stones, to provide a singular vista from the top. Foote, in contrast, has graduated to a sort of grassroots view, surveying the state of the natural economy through its trickle-down effects. And nowhere but down here, at the lowest trophic levels, do the animate and the inanimate come so close to converging: geology melts into soil chemistry; soil chemistry merges with biochemistry; biochemistry begets biology. There is a river of nutrients flowing up and down the scales of perspective. Foote had led me to the proverbial head of the Nile, to a biogeochemical spring bubbling out of the earth.

The properties of this spring, Foote believes, offer a potentially powerful measure of the impact that alien species have on the landscape through which the nutrients flow. There are various terms for introduced plants and animals: exotic species, nonindigenous species, invasives, nonnatives, aliens. "Nutrient additions" is the soil ecologist's way of saying it. Picture an ecosystem as a reservoir of nutrients, a bank of recycling, edible currency. The plants and organisms in the ecosystem are the bank's borrowers and lenders, the temporary holders of the currency. An *Astelia* lily sprouts from the forest floor, accumulating the nutrients necessary for its own growth; the lily decomposes, the nutrients rejoin the general pool. A damselfly larvae hatches in a leaf axil: withdrawal. A honeycreeper defecates from a tree branch: deposit. Invading species function similarly except that their transactions are, to the local ecology, novel. They are the embezzlers and money launderers: diverting funds into foreign accounts for nonnative use and flooding the system with unexpected infusions of nutrient cash. That is the theory, anyway. The question is, do the aliens make a difference? Do their deposits and withdrawals have a measurable impact on the productivity of the native ecosystem—on the amount of living stuff that the ecosystem can generate? Are native micropods less abundant in soil tilled by feral pigs? Are nonnative weeds more likely to grow there? Those are the kinds of questions Foote and his colleague Peter Vitousek want to get at. They are not unlike a pair of financial analysts poring over a series of bank statements for signs of fiscal misconduct.

"There are a lot of ideas floating around about how soil communities are structured," Foote said. "And there are a lot of ideas about what factors contribute to biological diversity overall. One of the basic ones is that more productive areas tend to have higher biodiversity than less productive areas. But almost always when people reach that conclusion they're comparing apples and oranges. They're comparing areas that are very dissimilar in terms of their origination and development. The nice thing about Hawaii is that you can argue that areas that differ in productivity may be very similar in other respects—climate, patterns of rainfall, temperature, the parent material underlying the soil. So it's a very nice model system for looking at overall patterns of biological diversity and tying that into this whole question of nutrient development and limitation. Is community structure determined by bottom-up factors—the amount of litter available for food—or from the top down, by predators? That's really what I'm looking at."

11

Foote's experimental plot, when we finally reached it, offered little to distinguish itself. A barbed-wire fence extended out of sight in both directions. Evidently we were on the outside of the exclosure, along with the feral pigs, wherever they might be. Foote pointed to tree limbs here and there, where yellow plastic ribbons marked transects through the woods; small red flags on wire staves indicated spots where Foote or someone from his research team would come to collect soil samples. The forest on the opposite side of the fence—the inside—was otherwise visually identical to the forest on the near side.

For Charles Elton and many of his successors, biological invasions were a way to probe and characterize the way that ecological communities are assembled and held together. The ecosystem was studied as a sort of organic machine—a system—of semi-interchangeable parts, or, to borrow another analogy, a kind of corporate economy maintained by organisms with defined ecological jobs. By studying the arrival of foreign workers and the consequent displacements, the notion went, a scientist could figure out the overall corporate structure: the various job descriptions, the interoffice competition, the company bylaws, the glue of market success and longevity. Critical to this schema is the job itself, the ecological niche—a concept that has receded from meaning over the years with every new attempt to clarify it. Today, one can speak of a habitat niche (the range of habitats in which a species can and does occur) or a functional niche, the "role" or "place" of a species in a community—a notion further divisible into trophic niche (the relationship of the species to its food and enemies) and resource niche (which spans things utilized by the species, like nesting sites). As a conceptual tool, the niche has effectively dropped from the modern ecologist's belt. "Niche," Mark Williamson summarizes in his book *Biological Invasions*, "is useful in a preliminary, exploratory description, but becomes difficult to pin down in particular situations."

Whatever a niche is exactly, successful invasion, in Elton's schema, allegedly involves occupying an empty one. But, many biologists counter, what does it mean for a niche to be "empty"? If a niche is an ecological job that takes up food or resources, and such a job is going unfilled in an ecosystem, the community would show side effects; it would soon be overwhelmed by waste or unused food. So for such a job opening to exist yet not harm the community, by definition it must be a job that involves no interaction whatsoever with other community members —like one of those jobs the boss's kid fills on summer vacation, only less productive. But if that is the case, any invader entering this empty non-job would have no impact on the system—which clearly isn't what happens with many invaders. "If you take the view that there are no empty niches," Williamson writes, "the invasion of communities cannot involve occupying empty niches." Some ecologists contend that in saying that an invader occupies a vacant niche, what is meant is that an invader plays a new functional role in the community, not that it doesn't use resources previously used by other species. The brown tree snake fits this definition; in coming to Guam, it declared itself top predator of an ecosystem that for eons had run perfectly well without such a top executive. Under these terms, Williamson writes, successful invasion becomes a matter of always, often, or sometimes entering a niche that can be full, empty, partly full, or partly empty—terminology that begins to suit the term itself. Williamson concludes, "It is to some extent a matter of the meaning you want to put on the word 'empty.'"

At the very least, invasion biology has made it clear that the conflicts and interactions that transpire between species in an ecosystem are too fluid and dynamic to be meaningfully described by a static term like *niche.* To introduce niche theory is to propose a koan: Does an ecological niche exist before an invader arrives to fill it? Meditatively interesting, perhaps, but useless in forecasting. "We are still unable to recognize a vacant niche except by carrying out the tautological experiment of introducing a species and seeing if it becomes established," one biologist notes. Williamson adds, "The extent to which a niche is vacant is, in practice, a post hoc explanation. Post hoc explanations are neither intellectually satisfying nor much use in prediction."

Foote and Vitousek aim to sidestep this conceptual quagmire. At all four experimental plots they will measure forest productivity—the total mass of life that a given area produces—both outside the fence (where pigs are) and inside the fence (where pigs aren't), over time, to see if there's a detectable difference. They are ignoring the ecosystem's corporate hierarchy—workers, job descriptions, and competion—to focus instead on the common currency that flows through the office. Rather than consider an ecosystem as an assemblage of parts, they treat it as a black box. What matters most isn't what's inside; it's what goes in and what comes out.

Quantifying the productivity of a forest is no small task, however. Vitousek, an expert on "ecosystem properties"—soil fertility, nutrient cycles, and the like—will concentrate on the soil and the flora that arises from it. How productive is Hawaiian soil under various conditions? What amount of green biomass—the dry weight of plant life—can a given area support? Leaf litter is gathered from selected areas, dried out, ground up, and weighed, and its chemistry is analyzed. How much litter accumulates over time? How much carbon, nitrogen, calcium, magnesium, and potassium does it contain? What percent of the nitrogen and phosphorus is actually available to plants, and what percent is inaccessibly bound up in compounds? At what rate is nitrogen released from organic matter decaying in the soil? As soils breathe and decompose, they release carbon dioxide, methane, and nitrous oxide—these gases will be measured over time too. Slim belts of metal, called dendrometer bands, will be placed around the trunks of selected o'hia trees to monitor the rate at which the trunks widen and the trees grow. Leaves will be gathered from the canopy, either with a shotgun or, near the state prison on the Big Island, a slingshot. What is the biomass of leaves in a given unit area? What is their nutrient content?

Those are Vitousek's concerns. Atop these, Foote is adding his own layer of inquiry, the soil dwellers: springtails, mites, and other microarthropods, extracted from the soil using Christmas lights and Berlese-Tullgren funnels; dried out; ground up; weighed; the biomass of the invertebrate population measured and tracked. And Foote will monitor the flies—the drosophilas, which complete their life cycle on decaying plants and bacteria and fungi, and the *Dolichopodidae*, or long-legged flies, which prey on the springtails and mites—which are as sensitive to shifts in their prey base as the honeycreepers are to the abundance or disappearance of their favorite seeds and caterpillars.

"The idea is to integrate it all together and try to understand the general pattern of soil alteration associated with pig digging and disturbance," Foote said. By creating a kind of data sandwich, he and Vitousek will be able to quantify the biomass and diversity of the invertebrates and compare that with the productivity of the underlying trophic level—the trees, plants, leaves, leaf litter, and the soil itself. "There should be a fairly close correspondence between the two," Foote said. One could take the comparison a step further, he added, and look at the productivity overhead, of plant-dependent invertebrates that live in the canopy. Beyond that, one could even examine the highest trophic level, the forest birds. How do changes in nutrient flow through the other trophic levels affect the birds, which mostly feed on insects or nectar from the trees? "That question is probably the least tangible, in terms of addressing it in any experimental way. But certainly with the lowest trophic level it'll be possible to look very closely at where there have been nutrient additions to see whether or not that affects diversity at all. And my guess is it probably does."

He added, "There's not going to be a simple equation that says pigs encourage weedy soil fauna. People have been introducing soil fauna here since the seventeen hundreds. There's been a tremendous transport of potted plants, agricultural plants—they must have had soil mites and other critters. When people talk about restoring ecosystems here to a pristine state, most of the organisms responsible for maintaining that system are foreign. The most important is the earthworm—and there's no evidence that they are native. They are all introduced species."

Foote had wandered away from the experimental plot, down one of the side trails through the forest. Here and there on the path, shreds of greenery were strewn about, the remains of an ieie vine or a lily fed upon by a feral pig. Foote followed from one to the next like a meteorologist on the path of a tornado. "I started finding earthworms in the leaf axils of the *Freycinetia* and the *Astelia*," he said. "It occurred to me that one reason why the pigs may be ripping off the leaves is simply to forage for the very small but maybe significant amount of protein matter that is tied up in these little leaf axils. There are always these garlic snails—an exotic snail—in there. And there are a fair number of earthworms. There's no easy way to test the idea; but I'm paying more attention to the way in which the pigs feed on the *Astelia* and *Freycinetia.* It really doesn't look like they're too concerned about eating the plant itself. But they sure rip 'em apart. I can picture them sticking their little tongues in, licking up isopods and worms, and leaving the foliage."

The afternoon was getting on. Foote followed the path of blasted lilies back to the main trail, then turned in the direction of the bog. A stiff breeze swept up from the canyon. The ridge opposite, normally visible through the trees, was cloaked in swiftly moving clouds. There was one task yet to complete before hiking to the truck: collecting flies. We were approaching the series of sponges Foote had tacked up a few hours earlier; even from a distance one could see that they were freckled with drosophilas.

"We're getting flies," Foote said. "One thing about drosophilas is, they're slow; they can be caught pretty easily."

In addition to the sponges and the mushroom-banana bait, there are two essential tools of the drosophilist's trade. The first is an aspirator, a three-foot length of slender, transparent tubing that Foote carries draped around his neck. With one end in his hand and the other in his mouth, he creeps toward a sponge, spots a particularly compelling picture-wing—"That's a really interesting one; I'm not sure what it is"—waggles a finger off to one side to distract it, then sucks the fly into the aspirator. A filter at the mouth end keeps him from swallowing it. Then with a gentle puff he blows the fly into a specimen vial, which he stops shut with the other essential tool, the cotton wad. Hawaii's more traditional drosophilists—Hampton Carson, Ken Kaneshiro, Bill Mull—cut their stoppers from old aloha shirts, a ritual that allegedly began when a wife prohibited a husband from wearing his aloha shirts in public once they became worn at the edges. Foote sticks to white cotton or cotton balls. "It's not Hawaiian," he said. "You can tell I'm from the mainland."

One virtue of chasing flies as a way to study pigs is the low likelihood of encountering the latter. It was only a few days after Foote and I returned from Molokai that I had found myself on the pig hunt with Lloyd Case, following a trail along the old Pu'u O 'Umi fence line in the Kohala highlands, on the prowl for something considerably swifter, larger, and smarter than any drosophila. In fact, I had fallen slightly behind. Case was partway up a low ridge when a bark sounded over the trees from nearby and he stopped in his tracks to listen.

He heard nothing: stillness, an exhalation of wind through branches. Case remembered a signature hoofprint he had seen a hundred yards earlier on the trail. "That track down there was from this morning," he said. Just then one of the dogs, Roxie, burst through the bracken onto the trail, panting, then dashed back into the forest. Spotty was absent. Case stood on alert, his rifle in his right hand and a machete in the other. Then we heard it: two more barks from close by, followed by a high-pitched squeal, and then a high, chilling yowl, almost human, like the yelp of a dog whose tail has been stepped on, only far worse. Concern flashed across Case's face. "The pig really got 'em, the dog." He paused to listen for a moment, then pointed up the trail with his rifle.

"The pig gonna come out on top there."

Nothing appeared. Case waited a few moments then stepped off the trail and began hacking his way through the forest with his machete. The ground was uneven, dropping sharply over roots, into pockets of moss, into mud. Roxie materialized from nowhere, running alongside us. Case yelled out Spotty's name. Again no response. The trees parted onto a clearing with a swath of matted grass. I wondered briefly what Foote might say about it, what a quantitative analysis might reveal about the pigs' impact in this place, but I did not linger on the thought. "Boars hang out in here," Case said. "I tell you, they catch a dog in here, rip him to pieces. Sometimes when you hear 'em like that, the pig kill 'em, that quick. Not many dogs cry like that."

We moved on. "I've lost maybe five dogs in my thirty-eight years of hunting. Sometimes the boar, he knows you're coming, and he'll stay right there, waiting. Just chop 'em to pieces. One boar claimed eight or nine dogs. One was amazing: he cut the bone and everything, right through the neck. Just a little bit of skin was holding it. That was one of my friend's dogs." He told of a boar out on Parker Ranch that would run up underneath a man's horse and slit the belly open. "Up in these mountains, you gotta be real careful. You get some monsters up here, five hundred pounds." Case let go a series of sharp, shrill whistles. "Hey, Spotty!" He handed me the machete. It seemed absurdly insufficient to its alleged task. I wondered if boars were like sharks, whether they could be dissuaded with a poke in the eye. I wondered, if events made it necessary, whether a pen would prove mightier than a sword.

Case plunged back into the forest and circled wide. The trees grew thick and close around us. It seemed easily possible to become lost. Case found a trail of fist-size droppings and followed it to another clearing, this one larger than the last and littered with fresh droppings. The air was acrid with animal pong. "This is the nest," Case said. "Looks like a lot of 'em, maybe ten or fifteen." We were in a wilderness labyrinth stalking a four-legged metaphor. But a metaphor for what? Captain Cook and his ecological imperialism? The quintessence of Hawaiian culture? Whatever it was, it had struck back. Case scanned the area for signs of blood but found none. Finally he sidled up to a tree near the center of the nesting area, where the scent of pig was particularly pungent. I heard a fly unzip and the trickling release of a bladder.

Case hacked the way back to the trail; I stumbled after. The clouds were low and glowering. Case whistled again for Spotty and shouted his name. The reply was a brisk wind and the soft clatter of leaves. It was a five-mile hike back to the truck, through shoulder-high bracken and grass. Case sized up the sky, which had taken on a grim cast, and began trudging up the ridge.

"We're gonna be in for some bad weather, boy."

As much as anything, a fence marks an intellectual boundary. On the one side, what is understood; on the other, what has yet to be grasped. Known, unknown. In an ideal world, the boundary is fairly clean and clear, and the bounded area of the known expands over time to enclose a yet greater portion of unknown territory. Foote has had the opposite experience. With regard to feral pigs, between what is known with certainty and what the various human factions believe they know with certainty, there is a remarkably wide gray zone in which very little of anything is known. The more Foote learns, the murkier it gets.

"In Hawai'i Volcanoes National Park we tend to think that pigs are responsible for disturbing the soil, reducing tree-fern densities, and then bringing in a lot of weedy species," he said on Molokai. "Well, there are pigs here"—he waved his hand in the direction of the trees outside his fence—"but I don't see any weeds. These are all native plants. So what else is in the equation? Well, in Hawai'i Volcanoes you've got a national park situated around a residential area where there have been huge numbers of plant introductions. So the opportunity for invasion is there, and pigs probably do speed it along. But whether banana poka and Himalayan raspberry would be there anyway is a very open question. We know that with Himalayan raspberry, birds feed very readily on the fruits; it's possible that the birds are also feeding on banana poka. Pigs may accelerate weedy species invasions, but on the other hand they may not. And there's no evidence to demonstrate it one way or the other. So it boils down to meetings with conservation biologists, land managers, and hunters, and you have three different opinions and no way to distinguish between them.

"At the very least, what we're trying to do here is quantify what we mean by pig disturbance. The same goes for documenting impacts on native vegetation: you've got to go out there and measure it. And frankly, when you do measure it, the effects aren't nearly as dramatic as the conservationists would have you believe. It's almost like a religious thing. I don't know how else to describe it: there is this tremendous belief that pigs are responsible for wiping out most of the rare plants. Except a) we usually don't have any information to show that those plants occurred at higher densities anywhere in the historical past; and b) we don't have any information about the historical densities of pigs. We can't even say that the pig densities were at a given level before the exclosure went up, because nobody went out and measured pig disturbance in a way that would allow you to see what has happened to this forest over time. The data that we're working on right now is some of the most comprehensive data available. I think the bottom line is that when the survey work is completed and we start to really quantify the impact of pigs, it's not going to be dramatic."

I couldn't help remarking that several hundred thousand dollars, the park's annual budget for its fencing and pig-eradication efforts, seemed like a lot to spend based on what sounded like a weak case.

"Well," Foote said, "I think that where eradication is practical, it's a first step toward a long-term stable recovery of the forests. But if you have an area that's already heavily invaded by alien plants and you don't have a mechanism for controlling them, and you have an area that has high densities of rats—and the densities of rats in our rain forests are among the highest ever observed in the world—then I wouldn't necessarily say that getting rid of pigs gets you anywhere fast. But if you have a method for controlling rats and alien plants, then I think that there's a lot to be said for clearing out the pigs. I don't want to stand in the way of people building exclosures. I think it's a good thing to do. But I think the hunters have a lot of good ideas too. We should be out there trying to test some of them."

One theory, expressed to Foote by a local hunter, is that removing pigs from open woodlands serves only to increase the density of the rat population; the grasses grow taller, the rats take refuge from owls and other predators, their numbers soar, and they ravage the succulent native plants. "It's perfectly likely," Foote said. "Some of my colleagues don't want to hear it. They think it's blasphemy. So they've been working with hunters' groups for years now and there hasn't been any improvement in the situation. Of course, I'm not optimistic that a lot of the hunters won't be real jerks about it."

Case and I returned home in late afternoon. He showed me his backyard and the enormous boar he kept penned there. I was exhausted. My clothes were soaked and streaked with mud. I wanted to be clean. I wanted everything to be clean. Some days earlier, while driving back from a meeting of biologists, state officials, and local hunters, one of Foote's colleagues had speculated that a real ability to preserve Hawaiian ecosystems will come when the community of native Hawaiian people begins to recognize and embrace what is truly endemic in the Hawaiian landscape. At the time, the words offered solace, a glimmer of optimism. Now I no longer knew what they meant. I grasped then how an ecologist—how anyone—could burn out in that environment, how faith in one's own views could acquire more weight than any evidence behind them.

Foote had offered me his porch to sleep on that night. I made the two-hour drive from Waimea to Volcano in falling darkness, struggling not to nod off. Foote works a farmer's hours: he is asleep most nights by eight o'clock and awake and in the office most mornings by four. He was still up when I got back, so we sat outside for a while drinking schnapps and "talking story," the local phrase for chewing the fat. The night air was dewy and reverberated with the pneumatic trill of a grasshopper. From somewhere down the street came the antiphonal barking of a neighbor's dogs. I asked Foote if his work ever discouraged him.

"What you get discouraged about isn't the loss of a few endangered species," he responded. "It's the fact that we haven't learned how to conserve very well. That we haven't learned how to share resources equitably. That humans are still at war with each other. That's far more critical. While human beings are still on the planet, we can still preserve reasonably good chunks of forest, provided you don't have a situation like Rwanda." He added, "You keep from being discouraged by focusing on those bits of land that have been set aside, then focusing on small chunks within that habitat. In the end, you can spend the same amount of money and save one species or hundreds. Obviously, if you have the money, you do both. It boils down to triage, essentially. If you have an endangered bird down to a few individuals, why waste time on it? I don't think people really disagree with that view, except that they think that focusing on endangered species is a useful way of preserving habitat. The next phase in conservation is going to be learning to accept some losses. We haven't really reached it until we're willing to do triage, and nobody is willing to do that yet."

The morning dawned cool and bright, an unusual turn in cloud-shrouded Volcano. It was Sunday. At Foote's suggestion, we packed Sam into the car seat and drove just over the crest of the hill to Kipuka Ki, a semi-open scrubland on the dry side of the park. We parked and walked a few yards down the road; Sam stayed in the car, asleep in his chair, to the sound of talk radio. The scenery surrounding us was little more than four hundred years old, the thriving postlogue of a lava flow that had swept through the area centuries earlier and wiped the slate clean. It was the first place Foote had shown me when I initially visited the Big Island several years before. "It may be just a postage-stamp–size reserve, but in ten years you won't see many alien elements here, to the point where they're not affecting the regeneration of the forest. A self-sustaining ecosystem: I think it's realistically optimistic. It's enough to make anyone happy about what's going on in Hawaii."

It had taken me weeks to grasp the importance of Foote's research—less the science of it, which was complicated enough; more its potential resonance in the human community, if the evidence of the pig's disruptiveness turned out to be as thin as his early results seemed to indicate. I wondered aloud if Foote felt an obligation to enter the larger dialogue, to perhaps steer the hunters and land managers to a middle ground.

"I'm not the person to do that," he said. "There are other people who have to be that vehicle for communication, people who won't do it but could: local guys who have a good rapport with hunters, who grew up on this island and have a lot of respect in the community. I'm hoping that if I amass enough information, somebody might be willing to take that information forward in five or ten years and say, you know, 'Here's some really solid data showing that pig disturbance in different areas really depends on the density of pigs.' If we can show that an area like Kamakou Preserve is relatively intact despite the long-term history of pig disturbance, that suggests that we don't need to put up any more fences. Hearing that would really make hunters happy.

"But it can't come from me. I'm the haole scientist. I don't know if you've been to any of those meetings but they just . . . It's just so easy to nail you for being an outsider. I keep a low profile. There are a lot of people who'll be listened to as people rather than as Great White Scientists from the mainland—which I'm afraid is always the label I'll be given. I think of myself as a producer of information. If they need it, they can take it. But if I run into people in the forest, I won't shy away from them. If a hunter wants to talk story, I'll talk story."

12

In Volcano, in the center of Foote's front lawn, a native hapu'u tree fern stands a dozen feet tall, a static geyser of greenery erupting from the grass. At the base of the tree, leading into the ground, is a hole—the entrance to a wasp nest. It materialized there one day some months back and Foote convinced his wife to let it remain as a kind of natural experiment, as perhaps it is for the wasps themselves: they are an alien species in Hawaii, *Vespula pennsylvanica*, a variety of yellow jacket native to the U.S. mainland. *Vespula* first arrived in Hawaii sometime in the 1970s, purportedly as an intact colony in a shipment of Christmas trees imported from the Pacific Northwest. That explanation has never quite made sense to Foote, as yellow-jacket queens typically do not form new colonies at that time of year. "But it takes an imagination to come up with alternative explanations."

It was Thanksgiving. We stood on the veranda, which Foote had recently built around the front of the house, and looked on as a steady stream of yellow jackets issued from the ground. However the species arrived, *Vespula* now ranks among the most disruptive insects to have invaded Hawaii thus far. Of Hawaii's six hundred fifty indigenous species of bee and wasp, none are social insects: they lead solitary lives, singly hunting the larvae and eggs of other insects and laying their eggs in isolated burrows. *Vespula*, in contrast, is social to a totalitarian extreme. Its members aggregate in highly organized colonies, constructing their paper nests in tree hollows, in underground cavities, even on the exposed crags of lava fields. They are aggressive, efficient hunters. Foote has watched them pluck caterpillars from the blossoms of trees. Left to its devices in a laboratory setting, a *Vespula* worker will ritualistically slice off the wings of a native fruit fly, then suck the juice from the carcass. In the wild, a colony of *Vespula* will scour the terrain as thoroughly as a pack of accountants on a corporate audit.

The species has spread through Hawaii with phenomenal success—a success, entomologists note, that corresponds with alarming declines in the numbers of several native insect species, which in turn pose a threat to the native birds that feed on the insects. Foote suspects that the dwindling populations of some drosophila species, which he once attributed to feral pigs and their impact on the local flora, may be equally due to predation by the yellow jackets. Native caterpillars of the genus *Cydia*, a favored food of the palila—an endangered bird that now resides only on the upper slopes of Mauna Kea—have been hard-hit by the yellow jackets.

"You can pretty much count off the insectivorous birds," Foote said. "Each one of these workers is carrying one or two masticated insects. The traffic rate is not a very reliable count: on a sunny day, you could see ten times that number. In New Zealand, they have these nests every ten meters or so."

As we watched, yellow jackets swooped in and out of the hole in his lawn at a steady pace, a new worker coming or going every second or so. I stood riveted. Here was a river of living particles, winged automatons, flowing in both directions at once with a regular, almost mechanical precision, yet with a fluidity suggesting some rudimentary, inchoate intelligence. I might have been watching a flow of electrical current or an exchange of bits through a modem cable. I felt I was witness to a dialogue of sorts, a conversation between the poles of inanimacy—this hole in the earth, the wide-open sky—written in the alphabet of *Vespula*.

What difference does one individual make? How does one organism or species contribute to the integrity of the community to which it belongs? Does its presence, or absence, matter?

It was Elton's hope that the study of invasions would provide some insight. He was particularly keen to investigate the nature of competition. As he rode through the countryside on his motorcycle, he saw a landscape of species busily fulfilling the various ecological tasks natural selection had assigned them, while simultaneously struggling to hold their ground against alien species that threatened to do the same job only better. Competition was itself a selective force and, Elton believed, contributed to the stability and biotic resistance of the encompassing ecosystem. The better adapted species were to their niches—the more honed by competition—the harder it would be for alien species to enter their midst.

However, competition was, and remains, vexingly difficult to demonstrate in the wild. The challenge is partly semantic: if one cannot precisely and consistently define the niche over which two species allegedly compete, it becomes hard to define exactly what their competition consists of. Historically, the phenomenon has been more readily observed among plants. Plants don't move, so the resources over which two plant species could be said to compete—sunlight, soil nutrients, water—are easier to monitor and measure, both outdoors and in the laboratory. Animals are not so static, however, and the range of resources over which any two species of fish or fowl might tangle is much wider, including mates, breeding grounds, and, for an omnivore, food of all types. (Unlike plants, animals also engage in predation: they eat or are eaten, which in either case is an additional pressure beyond competition.) Elton was painfully aware of these limitations. "We do not get any clear conception of the exact way in which one species replaces another," he noted in *Animal Ecology*. "Does it drive the other one out by competition? And if so, what precisely do we mean by competition? Or do changing conditions destroy or drive out the first arrival, making thereby an empty niche for another animal which quietly replaces it without ever becoming 'red in tooth and claw' at all?" Even Gause's classic experiment, involving two species of paramecium and some bacteria (their shared food) in a test tube, was less than definitive in illustrating competition. Gause found that as the experiment progressed, the animals changed their environment: their waste accumulated, which altered the animals' growth rates sufficiently that one species consistently beat out the other. Clearly, poisoning your opponent with your feces is an effective strategy for microcosmic domination; whether it fairly qualilfies as competition is another matter.

Elton was confounded in particular by the case of the American gray squirrel. The squirrel was introduced to England in 1876, perhaps for hunting purposes or purely for visual variety; thanks in part to some thirty subsequent introduction efforts, it is now widespread in that country. Innocuous enough in its home range, the gray squirrel—"tree rat" to its detractors—has made a pest of itself in its new land with its habit of eating flower bulbs and birds' eggs and stripping the bark from young trees. In addition, the spread of the gray squirrel has coincided with declining numbers of the Eurasian red squirrel, a beloved native despite the fact that it is only slightly less destructive than the gray squirrel.

Outwardly, Elton wrote, the situation bears all the signs of an invading species outcompeting a native one. Both species of squirrel eat acorns, and, well, how ecologically different can two acorn-eating squirrels be? But, Elton continued, there was no hard evidence that competition explained the red squirrel's decline. "There are thousands of similar cases cropping up," he added, "practically all of which are as little accounted for as that of the squirrels." Even today, with fewer than twenty thousand red squirrels remaining in England, the precise cause remains unclear. Scientists have found that the gray is more efficient at foraging in the woods and in backyards. On the other hand, even before the gray's arrival, red squirrel populations in Britain had a frequent tendency to die out. (They were reintroduced to Scotland and Ireland several times during the nineteenth century.) In addition, it is now known that two-thirds of gray squirrels are silent carriers of a viral skin disease fatal to red squirrels. Again, domination comes easier to those who can spread a pox. But is that competition? In any event, between the virus and the concerted human effort to introduce gray squirrels, it is hard to credit competition singly, or even chiefly, for the success of the squirrel's invasion.

Lately some biologists have pointed out that many successful invasions that have been attributed to competition can be explained through a far simpler mechanism: opportunity. The gray squirrel thrives in England in no small part because humans intervened repeatedly on its behalf. The English house sparrow failed to take hold on American soil when eight pairs were initially released in Central Park in 1851, a failure that could be ascribed to any number of reasonable causes, including competition with native birds. But the English sparrow did catch on the following year—almost certainly, Mark Williamson notes in *Biological Invasions*, because this time *fifty* pairs were released, followed by other releases elsewhere in the country. "It is all too easy to think there is competition where there is not, or to think there is not when there is." Stuart Pimm of Duke University and Daniel Simberloff of the University of Tennessee in Knoxville, have disagreed for years over the real reason for the success of certain foreign bird species now resident in Hawaii. Pimm has argued for the birds' competitive superiority. Simberloff instead notes their anthropogenic history: they succeeded because so many nostalgic bird-club members worked hard to ensure their successful introduction.

In fact, it may be that the classical concepts in ecology—competition, niches, species richness, community structure—have little bearing on why any given invader succeeds. Rather, an increasing number of invasion biologists are coming to the conclusion that the main factor of invasion success is nothing more complicated than propagule pressure:

the frequency and persistence of the introduction. Throw enough different kinds of noodles at a wall for long enough, and eventually one will stick. Agricultural plots, roadsides, vacant lots, and other landscapes disturbed by people do see more invasions than undisturbed areas, but not because they are less rich in their number of species. Rather, Williamson counters, "it only reflects the fact that species are more likely both to be transported from disturbed areas and to arrive in them, because of human activities." The statement that disturbed areas are more easily invaded is essentially tautological: disturbed areas are more invaded, so they have more invaders. Propagule pressure likewise explains Hawaii's general plight, argues Simberloff: the lowlands harbor more introduced species than upland forests not because the lowlands are disturbed or simple or vulnerable, but because human traffic is heaviest there, and heavy traffic creates more opportunities for an invading organism to succeed. What to Elton's eye looked like biotic resistance merely represents, to Simberloff, "the historical distribution of dispersal opportunities." As Williamson puts it, "The number establishing reflects the ease of transport."

And it isn't only humans that transport and disperse introduced species. In the eager search for evidence of competition, Simberloff writes, biologists have overlooked the extent to which introduced species mutually abet one another's success. The lantana vine was an innocuous introduction in Hawaii until the introduced mynah bird began to eat the vine's fruit and disperse the seeds in more distant forests. The spread of avian malaria among the honeycreepers represents the confluence of three invaders: the *Plasmodium* parasite; introduced songbirds, which carried the parasite in their blood; and a mosquito vector. It's all about opportunity. Humans are often the initial agents of dispersion, but the seeds we spread also spread seeds in turn: propagule pressure squared.

In an Eltonian world, the probability that any given invasion will succeed should drop over time as the number of introductions rises: more species now living in the ecosystem, more competition for niches, lower odds of success for later entrants. Simberloff proposes precisely the opposite argument. If introduced species help one another rather than interfere, then the odds of an invasion succeeding should increase with the number of invasions that preceded it. Invasion begets invasion. "It is possible to imagine an invasion model very different from the dominant scenario of biotic resistance," Simberloff writes. In the extreme, one can imagine a string of begats, an "accelerating accumulation of introduced species and effects." Simberloff calls it "invasional meltdown."

In a geological sense, the Big Island is only the latest manifestation of an ages-old, ongoing meltdown. From a volcanic vent in the seabed, magma oozes up, erupts, and at last after eons blossoms above the waves: an island, the Big Island, the youngest island in the archipelago called Hawaii. Today the fountainhead is Kilauea Crater; before that it was Mauna Loa proper, and before that Mauna Kea, and so on northwest across the island, across all the islands, across and through the entire sunken chain. The vent may sit dormant for thousands of years. Since 1983, Kilauea has erupted almost daily. Its issue creeps downhill in burning streams and rivers, carves through the forest, flattens some tracts, and leaves others unscathed. When the lava comes, all biological accounts are wiped clean: the ground is charred and buried, time begins again at zero, and before long the bank of nutrients reopens to any and all borrowers. Every fresh lava field is a new invitation for life—plant or animal, native or not—to try and settle in.

One afternoon, I drove with Foote to the dry side of Hawai'i Volcanoes National Park, down a long, gentle slope on a lower flank of Kilauea, to a site where a new, biological meltdown was occurring atop the old, geological one. The terrain was arid and open, a buckled field of lava from which, here and there, young o'hia trees rose up: gray, gnarled trunks, and branches with red blossoms like small explosions of bergamot. Everywhere else, grass—chest-high and sun-dried golden—stirred in a breeze that drifted up from the sea, which was visible several miles in the distance. "When Kilauea erupted in 1959, it dumped a huge amount of ash and cinder throughout here," Foote was saying. The fragrance of the grasses that have grown up since was intoxicating, dewy and sweet like something from the Serengeti Plain. It was all alien, Foote said. Some was beard grass from Central and South America; some was broom sedge from the southeastern United States; some was molasses grass from Africa. All three were introduced to Hawaii in the early 1900s as cattle feed for ranches throughout the islands. The wind-borne seeds soon jumped their confines, and a habitat novel to Hawaii—the grassland—has begun appearing across the islands, its spread uncontrollable. The 1959 eruption was their opportunity to invade the park.

Normally, a bare lava field in Hawaii would grow over to form an open woodland dominated by native o'hia trees and mamane shrubs. Instead, the alien grasses, quick to settle on open ground, grow in thick mats that shade out most other plants, favoring only themselves. The novel appearance of grasslands in turn has introduced an entirely new force of nature—fire—to Hawaii. Since 1968, with a new and persistent bed of fuel to burn, the number of wildfires in the Hawaiian Islands has jumped fourfold, and each fire on average is sixty times larger. Already, several areas of the park targeted for restoration have been razed by wildfires. The invading grasses are ideally suited to the situation: their roots can withstand high temperatures, and their seeds, dispersed by the flame-fanned winds, are quick to sprout in charred soil. Various studies around the park have found that the total area covered by alien grass increases after a fire and remains thus dominated even two decades later. Every fire sets the stage for another, more destructive, hotter fire—and further invasions of fire-promoting grasses. If an "invasional meltdown" involves one invader dispersing another, this was a meltdown by fire.

According to Simberloff, one introduced species can abet other invaders not only by dispersing them but also by altering the environment in a way that increases the odds of their survival. Foote's colleague Peter Vitousek has made a specialty of studying such invaders: species that "not merely compete with or consume native species" but instead alter "the fundamental rules of existence for all organisms in an area." Melaleuca, an Australian tree that is spreading rapidly through the Florida Everglades, draws in so much water through its roots that it essentially converts open marsh habitats and wet prairies into shady forests or dry land—and leaves the marsh dwellers homeless. Like the alien grasses in Hawaii, melaleuca greatly increases the risk of fire, which clears the land and spreads the tree's seeds. Many floating aquatic plants can be thought of as rule-changers, insofar as they can dramatically reduce the amount of sunlight that filters down through a previously open water column. Goats and cattle change the ecological rules of islands like Hawaii by eating and trampling vegetation and causing the soil to erode. Whether the feral pig, by burrowing into the soil of Hawaii's rain forests, is a rule-changer too is part of what Vitousek and Foote hope to learn with their fencing experiments on Molokai and the Big Island.

In essence, a rule-changing invader alters the flow of nutrients through the landscape in a manner that benefits itself and often, incidentally, any number of other invading species. On this day, Foote had driven out to check on the advance of one such intruder, a tree—*Myrica faya,* more comonly known as fire tree—that lately has begun to encroach on his research into both feral pigs and soil micropods. The fire tree earns its name not for any fire-promoting qualities, but rather for the clusters of small orange blossoms that decorate its branches. The name has proven apt, however, as *Myrica faya* has shown a tendency to spread with burning speed. Originally from the Azores, the fire tree came to the Hawaiian Islands in the late 1800s, probably as an ornamental plant. It gained notice in Hawai'i Volcanoes National Park in 1961; by 1977, roughly fifteen hundred acres of the park were occupied by fire trees. Within another eight years the plant had spread to cover nearly thirty thousand acres of park and nearly ninety thousand acres in the islands as a whole. It tends to thrive on young soil—volcanic cinder less than fifteen years old—and in semi-open forests. Its spread is aided by birds, mostly introduced ones, that feed on its berries, fly off, perch on native trees, and deposit the seeds directly below. But it was the 1959 Kilauea eruption that opened the door widest, Foote said. "That disturbance allowed for *Myrica faya*, which was already well established in the park, to really spread. As we drive down the hill, you'll start to pick it up."

The truck clattered over ruts in the road. Gradually the neo-African veld of introduced grasses gave way to a more wooded area, where Foote pulled over. With instruction, a novice tree-watcher could distinguish between the native o'hias, with rounded leaves of silver-gray, and the darker foliage of the fire trees. "If you look across the landscape, you'll see trees with pointy crowns, a bit like Christmas trees, that contrast with the o'hia. Right here it looks like *Myrica* is occupying greater than about sixty percent of the ground cover. It's really just choking out the o'hia."

The secret to *Myrica faya*'s success, Foote explained, lay hidden below our feet. The plant is a nitrogen-fixer: its roots harbor bacteria that gather nitrogen, an essential nutrient, from the surrounding soil and pass it on to the plant. In Hawaii, this trait confers an enormous advantage. In any environment, the fertility of the soil is determined foremost by the amount of nitrogen it contains, more being better. Volcanic ash and lava, which form the soil base for most new ecosystems in Hawaii, are notably lacking in nitrogen. Most of the native plants that first sprout on a Hawaiian lava field are adapted to nitrogen deprivation; they grow for so many years until a layer of soil—the decomposed remains of the first generation of colonizing plants—accumulates and contains sufficient nitrogen that a new and different generation of plants, seedlings adapted to a more nitrogen-rich diet, can move in and begin to dominate the scenery.

Nitrogen-fixers—or more precisely, the specialized bacteria employed by the roots of nitrogen-fixing plants like the fire tree—are uncommon among Hawaii's flora.

"In a nitrogen-poor regime like this, any species that is capable of manufacturing its own nitrogen will have an advantage," Foote said. "And that seems to be what we're seeing here with *Myrica faya*. All of the native species here are classic colonizing species. They pop up on young lava flows and are perfectly capable of growth and development in a nitrogen-poor lava matrix. It's just that *Myrica* can grow faster and reproduce more rapidly."

Even as it hogs space aboveground, *Myrica faya* is radically changing the makeup of the soil below. In a series of careful experiments, Vitousek has found that the soil and forest floor beneath *Myrica faya* contain significantly more available nitrogen than do areas nearby. By making nitrogen so readily available, *Myrica faya* may be favoring the success of other nitrogen-hungry invaders that ordinarily would not be able to colonize fresh lava fields. New species, new rule: nitrogen. Moreover, the fire tree establishes and grows quickly enough on fresh lava fields that it soon shades out other, more light-hungry native plants trying to get a start there. Among the invaders profiting from the fire tree's nitrogen bounty, Foote said, are earthworms; their populations appear to be particularly dense in the nitrogen-rich leaf litter dropped by *Myrica faya*. Earthworms are soil movers and shifters. Foote has found that in areas where earthworms are numerous, the numbers of endemic micropods—the Hawaiian mites and springtails, which thrive in the undisturbed soil of old forests—are unusually reduced. In turn, the new inhabitants of the soil alter the manner in which nutrients flow through it, in effect altering the platform on which successive generations of life can build. Foote said, "Earthworms definitely have a very significant effect on rates of decomposition and nutrient cycling. So if you have an invasion of an alien tree, you have a subsequent increase in the density of an alien megainvertebrate that then alters the structure of the endemic invertebrate population in the soil. That's something of concern, but there's not much we can do about it." Earthworms, he added, also appear to be a snack of choice among feral pigs.

In its effect, the fire tree behaves something like a floodgate operator, sending a wave of nitrogen down what would otherwise be a moist creek bed; its downstream effect is less a meltdown than a washout. As the first settler to arrive on lava-bared terrain, it resurfaces the nutritional landscape for all colonists that would follow. This troubles an ecologist, but it is also interesting because it again raises the old question of "keystone species"—whether individual species can be said to carry out ecological roles; whether some roles matter more than others; whether certain native species are like rivets, more essential than others to the general health and integrity of an ecosystem. Remove them, and the rest of the ecosystem unravels around them. If an invader can do it—if a single species of tree like *Myrica faya* can fundamentally alter the shape and contents of an ecosystem—then it stands to reason that some native species must play a similar role by setting the rules for their own community, Vitousek notes. If some invaders "matter" to the future course of an ecosystem, then some native species must matter too. "We believe that a demonstration of widespread ecosystem-level consequences of biological invasions would constitute an explicit demonstration that species make a difference on the ecosystem level."

Still, Foote cautioned, it is possible to overstate the singular nature of *Myrica faya*'s impact on the local dynamic. The tree was certainly altering the local rules of nutrition, but the rules had been in flux for some time already; teasing out how much each new invader contributes to the overall change in scenery is nigh on impossible. Where he and I now stood, feral goats, set loose a century earlier to forage and multiply, had once roamed by the thousands, until park officials eradicated them in the 1970s. By that time, however, the goats had greatly reduced the abundance of various native plants, including the mamane shrub, one of the few Hawaiian plants that can fix nitrogen and typically an early settler on fresh lava fields. Had mamane been more prevalent, Foote said, *Myrica faya* might not have invaded so thoroughly. Paradoxically, in eliminating the goats, the park removed the one force capable of trimming back the alien grasses. Fewer goats, more alien grasses, a rising incidence of wildfires; more wildfires, even more alien grasses. The dense growth of the grasses in turn has made it difficult for mamane to reestablish itself, giving *Myrica faya* more room in which to spread out. And without the fleshy seedpods of mamane available to feed on, the endangered palila bird has retreated to the upper slopes of Mauna Kea. In some cases, it seems, the only thing more difficult than watching an invader transform itself into a keystone species is trying to extract it and reverse the damage.

13

On occasion, lava breaks through a wall of Kilauea Crater or issues from one of the smaller craters around it and makes a run for the sea. Often enough this happens in plain sight, and it is entirely unstoppable. Everything is plowed under: swaths of forest; stretches of pristine beach; even, as happened in 1989 along the Kalapana coast thirty miles downhill from Kilauea, expensive houses with ocean views. It is common enough these days to read about nature giving way to culture, of a stretch of marshland, say, lost as bulldozers and tar spreaders prepare the ground for a housing development or a chain bookstore. Rarely do you hear of the converse—civilization paved under to make way for some trees. Yet that is exactly what happened. The residents of Kalapana could only watch as the lava inched closer each day like an assault of red-hot snails: sending trees into flame, engulfing a park visitors' center, creeping across a million-dollar roadway, and one by one reducing homes to carbon ash. I watched it all years later on a local television station devoted exclusively to footage of advancing lava flows, as the narrator spoke ominously of the goddess Pele exacting her revenge.

The flows of lava harden and overlap in a tangle; they can be several miles wide, a hundred feet thick. On the wall of my Manhattan apartment is a map that traces the dates and jumbled outlines of individual flows: 1840, 1982, 1971, 1919, 1955, 1972, 1823, 1969, 1974, and one that has run continuously from 1983 to the present day. In person, however, to the untutored eye, the flows are indistinguishable, barren, disorienting—an undulating ink-dark glacier, an anti-glacier. Yet the flow of lava is not always aboveground nor visible. Consider a stream of lava—notably of the highly fluid, basaltic variety known as pahoehoe. It runs downhill, following the contours of the earth into gullies and natural channels. Gradually the sides of the stream cool, congeal, harden; the active lava runs now like a brisk river through a narrowing canyon. The surface of the river may cool and harden too, crusting into a thick roof. And still the lava advances, molten blood churning now through a roughly cylindrical tube several or a dozen feet belowground.

When the flow ceases and the lava drains, a tunnel remains, its walls fused to a black glaze by the combustion of volcanic gases. With new eruptions in subsequent months, years, or centuries, new tubes may intersect and infuse old ones or cut them short, crossing and recrossing, interweaving, forming an interrupted plexus of hardened, emptied arteries—interconnected further by restless capillary crevices that open and close as the lava field cools, cracks, settles, ages, and erodes. To venture into a lava tube, then, is to enter the circulatory system of the earth itself, or, it often seemed to me, the Jurassic pathway of some giant rock-eating earthworm. Some lava tubes extend for a hundred yards before breaking into skylight, others for a thousand. Some, like the Thurston Lava Tube, not far from the headquarters of Hawai'i Volcanoes National Park, are large enough to permit tourists to stroll through two abreast. Many more lace beneath long-overgrown lava fields in forgotten corners of Hawaii, and they are as cramped as a crawl space, or narrower—certainly no place, one might easily assume, for an explorer of Frank Howarth's bulk.

Howarth is an entomologist at the Bishop Museum in Honolulu, where he holds the L.A. Bishop Chair in Zoology. Although the position is distinguished, the accommodations are burrowlike. Howarth occupies the last in a row of makeshift cubicles, each separated from the next by walls of steel bookshelves stacked with dusty journals and jars of pallid creatures steeped in formaldehyde. He is large, with burly forearms laced with scratches earned in search of this endangered damselfly or that yellow-faced bee. A dense white beard merges into a shock of silver hair atop his head. His glasses—thick, with plastic rims—are affixed to his head with a wide rubber-band strap.

In the 1980s, Howarth was among the first to point out that the presence of introduced species seemed to foster further invasions, not foreclose them as Elton's competitive-niche paradigm suggested. He reached this conclusion in part by studying the history of biological control in Hawaii: the practice, mostly by agricultural interests, of introducing one species (most often an insect) to combat another one (usually a plant, but sometimes another insect) that has made a pest of itself. He is a leading scholar and outspoken critic of biocontrol, with a pointed question: If it is impossible to accurately predict what an organism will do when introduced to a new environment, and if, moreover, every new invasion opens up new opportunities for subsequent invasions, how can anyone purposefully introduce any organism and state confidently that it will stay contained? When the Polynesian voyaging canoe *Hawai'iloa* arrived from the Marquesas infested with biting midges, Howarth was one of the experts the Hawaii Department of Agriculture called for advice. For the television news, he called the insects "a potential disaster for Hawaii." In private, he jokingly refers to them as "biocontrol for tourists."

In my search for an undigested nature I had followed Foote into the porous realm of soil, to the decomposing, perpetually digested boundary between vegetable and mineral. It did not occur to me that there might be still more to see farther below, but there was. Howarth's entomological specialty is lava tubes and the strange array of insects that inhabit them: albino cave moths, albino plant hoppers, pale crickets, and earwigs—most of them blind, having evolved in near or total darkness. They are descended from insects often still living aboveground. The big-eyed hunting spider, found outdoors on mountain summits, in young lava flows, and in forests, has two cousins relegated to the caves: the small-eyed big-eyed hunting spider, and the no-eyed big-eyed hunting spider. Howarth has discovered more than fifty different species of lava-tube insects in Hawaii so far, although, as he quickly insists, he merely finds them. The real work of identifying a new species is performed by the taxonomist he often sends them to, whose task involves counting tiny hairs on tiny legs and deciphering even tinier genitalia.

Howarth is constantly exploring old caves and looking for new ones. He studies a map of the lava flows, locates a likely area, goes there, and stumbles around until he sees a promising vent in the earth. He straps on knee pads, slips on gloves, dons a hard hat with a miner's lamp. "Ah, wild caving at its best," he'll say, and crawl in. Typically, the floor is level and the ceiling is low. If Howarth has chosen well, delicate strands of roots hang from roof to floor, in curtains. The o'hias, the first native plants to colonize a fresh lava field, send their roots dozens of feet deep in search of water. The roots are the life-threads of the cave, the fundamental source of moisture and nutrients. The larvae of the cave moth and the nymphs of the plant hopper—a white fleck of an insect—feed primarily on them. The other inhabitants feed on these insects or their debris. It is an ephemeral environment. Older lava fields are dominated by koa, soapberry, and other trees, whose roots do not extend below the blankets of soil accumulating there. For this reason, cave-adapted insects are limited almost exclusively to younger lava tubes. In one five-hundred-year-old tube near Kilauea, Howarth found eleven species of cave-adapted arthropods and insects. In another, only a hundred years old, seven species had already settled. In fact, the insects rarely inhabit the caves proper: they spend most of their lives in the interstitial cracks and voids, venturing into a new cave only when the conditions—high temperature, high humidity, a stable air mass—are right. The trick, Howarth says, is to find a cave that enters their environment.

Their environment, by most standards, is hostile. In the cave Howarth agreed to show me, we could see our breath, not because the air was cold, but because it was warm and humid. The farther into a cave one ventures, the less the climate is affected by changes in the temperature and weather in the outside world. Some lava tubes contain what Howarth calls the deep-cave zone. The climate in these nether regions, reached only through tortuous crawlways, is virtually isolated from the rest of the cave; the air is stable and very warm, the humidity high and constant. Beyond the deep-cave zone, in the best tubes, lie pockets of stagnant air. The humidity here is 100 percent, the temperature more than a hundred degrees Fahrenheit. The concentrations of oxygen and carbon dioxide can fluctuate dramatically, depending on the amount of decomposing organic matter. The resident insects can cope with these conditions. An entomologist can't, not for long anyway.

"The air is saturated," Howarth said, "so your lungs don't work—you're pulling in water from the air, and pretty soon your lungs fill up and you drown. And you overheat—there's no way to cool off." A few years earlier, a local town sealed up a cave after a tourist entered, became lost, and died from heat prostration. On those occasions when Howarth reaches the deep cave, he does not linger. Fortunately, there are cave insects all along this climatic spectrum, forming a morphological spectrum of their own. Howarth pointed out a cave moth: large and gray, with numerous black scales and distinctive white marks on its wings. Out in the rain forest, he said, is a closely related species, identical but for bold black and white stripes on the wings. Farther into the cave, on the lip of total darkness, is yet another relative—fully gray with only pale wing patterns. And in the deep cave is yet another, its wings undecorated and, on the female, tiny and useless; the male of the species is blind.

Howarth crawls through a lava tube with the meticulous grace of a bear in a blueberry patch. He makes a point of not breaking roots, or—to the extent that this is possible—even brushing them. A plant hopper nymph, knocked from its edible perch, could well starve. Trampling a visible root can kill the fifty additional feet of it that continue down through the rock, thus removing it from the food chain. Accordingly, Howarth does not like the notion of visitors in the caves. People leave trash, write graffiti, and exhale cigarette smoke, which in enclosed spaces acts as a powerful insecticide. Even informed visitors draw his skepticism. Once, Howarth rounded a corner to discover that his companion, a local conservationist and astute field biologist, had written the word *shit* in a small patch of white cave slime. To Howarth's subsequent lecture, the fellow replied that he merely had been marking the location of a small pile of rat droppings occupied by some intriguing insects. "Both of his arguments were patently false, of course," Howarth later wrote in a paper on cave conservation, "and after some consideration I erased the word even though it virtually destroyed the remains of a rather nice patch of slime and its inhabitants." Howarth is presented with the ecological equivalent of the Heisenberg uncertainty principle. Publicizing the existence of the lava-tube ecosystem will surely attract curiosity seekers and bumbling explorers. Remaining silent ensures that the caves will continue to be paved over, chewed up by developers, destroyed in ignorance. Even in his scientific papers, Howarth is cagey: he describes roughly where his caves are, but never exactly.

Although Howarth loves all his lava-tube insects equally, he has a soft spot for the plant hoppers because, as he has discovered, they sing. He traces the roots of this insight back to 1949 and the work of the Swedish entomologist Frej Ossiannilsson. With a simple set of tools, Ossiannilsson demonstrated that some eighty local species of the suborder Auchenorhyncha—which includes leafhoppers, plant hoppers, and various other insects not much larger than the typewritten letter I—produce an astounding array of sounds. But you must listen closely. The first song Ossiannilsson heard was emitted by *Aphrodes bifasciatus*, a leafhopper: he had put the insect in a small glass vial and held the open end to his ear. "By this method I have heard most of the calls described in this paper, and several species caught during my travels I have been able to study in this way only." Sometimes, to facilitate observation, Ossiannilsson would watch in a mirror as his subjects sang into his ear. Other times he would wrap a length of wire around the vial, insert the wire's other end, also spirally wound, into a second vial, then place the closed end of this second vial against his ear; this way he could view the insects through a magnifying glass. (Ossiannilsson later graduated to a stethoscope.) Yet other calls were strong enough that they could be heard distinctly if the insect was placed on a suitable sounding board. In most cases he used for this purpose the belly of his violin.

It was as though Ossiannilsson had wandered into a microcosmic concert hall, a veritable subsonic symphony. Each species of insect had its own call, as unique as a songbird's. Ossiannilsson got hold of a microphone and amplifier and made gramophone recordings. From the Royal Telegraph Society he borrowed a radiograph to measure the frequency of the rhythms. Limited only by the written language, he set out to describe what he heard.

Dicranotropis hamata: "I listened to the sound-production of this species on 22 June, 1945, at 5:35 p.m. Cluckings pitched in about C^1, emitted singly or in an irregular rapid succession, are often heard. These calls sound like a plucking on rubber strings, or a rapping on hollow wood. Sometimes a dull 'booooh' or 'biiiir' pitched in G^1 are heard. The latter sound lasts for two or three seconds. I heard these calls even in the dark on 23 June, 1945, at five minutes past midnight. Dying yell of the male: a short buzz pitched in F^1."

Calligypona elegantula: "The song was rather faint. It began with a couple of short croaking notes 'cha-cha-' followed by a rhythmical buzzing as from a sewing machine. This song lasts for about ten seconds."

Aphrophora alni: "As a sign of apprehension or dissatisfaction this male emits a smacking sound usually repeated in rapid succession and at intervals of irregular length."

Through his binocular magnifier Ossiannilsson deciphered the musical mechanics of his subjects. Orthopteran insects—crickets, katydids, locusts, and grasshoppers—generate sound by rubbing the edge of one forewing against another or against a leg; they are the string section of the insect orchestra. Ossiannilsson's subjects operated differently. They produced sound by vibrating their abdomens, as a tin can might sound if it were rapidly scrunched and unscrunched. (Cicadas, larger cousins of the leafhoppers, sing in this way.) In addition, the vibration of the body was often accompanied by a regular stamping of the insects' feet. In this sense, their song is more properly a drumming; Ossiannilsson compared it to Morse code. And as he soon discerned, it is a code with a purpose.

With species after species, Ossiannilsson beheld complex and musically furious mating rituals. "During the singing the abdomen vibrated vertically," he noted after watching a pair of *Eupelix depressa* through his binocular magnifier. "Soon this vibration became rather strong and set the whole body in such agitated motion that the male was forced to shift his foothold once a second in pace with the music. In this way a sort of dance or marking time was executed, interrupted, however, now and then by short, rapid movements in the direction of the female. The latter often tried to escape by running or by a leap. Sometimes the male made a short pause in his music but recommenced it if he found himself near the female. Having mounted the female and being busy trying to connect his genitalia to hers, the male continued his singing. During copulation he kept silent. The act of copulation took in the case observed a time of 15 minutes. When it was over, the male again performed a short song of the same kind as that above described."

Howarth surmised that something similar might be going on among his own lava-tube plant hoppers. Sure enough, experiments that he and a colleague, Hannelore Hoch, conducted with a soundboard—not a violin, but a small balsa wood stick attached to a sensitive microphone and amplifier—revealed a rich suite of songs. The call of each species, they determined, is unique. (The cave-dwelling plant hopper *Oliarus lorettae* is named in honor of country star Loretta Lynn, on whose property, and in whose spelunking company, Howarth found it.) In fact, when Howarth and Hoch played back mating calls of the plant hopper *Oliarus polyphemus*, which is found in seven caves on the Big Island, they heard seven slightly different dialects of song, each native to one of the seven caves. Each population was musically isolated, a separate sonic island. They sounded like seven separate species, in effect—and indeed may be. After some detailed peeping, Howarth found distinct, if barely perceptible, differences between their genitalia.

To breed successfully, potential mates of the same species must first recognize each other. For insects, as for birds, a song is one means of attraction. For eyeless insects living in total darkness, the song is the only means. But there are acoustics to overcome. The cicada is audible because he is large: his song is sufficiently strong to vibrate air molecules for a mile around. Pity the tiny plant hopper and her faint tremblings, too weak to vibrate the air. Howarth wondered: How does the male hear her? And how, having heard her song through the dark, does he find her? In the 1970s it was shown that aboveground plant hoppers transmit their signals not through the air but through the ground—more precisely, through their host plant, the stems of grass they spend most of their lives standing on and feeding from. For the lava-tube plant hoppers, the medium is the roots of the native o'hia tree. Howarth has found that the songs of subterranean plant hoppers can travel through roots for more than two meters—about the distance, in human scale, of a crosstown phone call. Songs can even travel between roots, via the hair-size rootlets that cross and touch the main lines, without the message seriously degrading.

When Howarth enters a lava tube, then, he sees a web of telegraphy. The lines around him are inaudibly electric—thrumming, whirring, buzzing. He is Horton hearing a Who. And that, he says, is getting harder and harder to do, for the simple reason that the roots are disappearing. As the introduced tree *Myrica faya* invades the fresh lava fields aboveground, the native o'hia trees are losing ground—and the roots of *Myrica faya* do not penetrate the cave roof and dangle free, nor are they edible to plant hoppers. The meltdown has seeped even to here. The lines of song are fading. The singers are growing mute, each one marooned on the island of itself, unable to communicate, to mate, to sustain its end of the evolutionary conversation.

Yet no root, native or new, can count on permanence. After showing me the monotypic stands of *Myrica faya* overtaking the lava field aboveground, Foote drove on for another few hundred yards, then stopped, hopped out, and waded several yards off the road into chest-high grass. The footing was sharp and uneven, the signature of an old a'a flow. Chunky outcroppings of lava dotted the field like small black drumlins. Here and there a tree stood: mamane, red-blossomed o'hia, and specimen after specimen of *Myrica faya*, the latter looking less than fully healthy. Foote plucked a leaf from one nearby. Normally large and green, this leaf was shrunken, yellow, and cupped in the manner of a canoe or a honeycreeper's tongue. "Foliar chlorosis," Foote said. Recently biologists at Hawai'i Volcanoes National Park had begun to notice an ailment afflicting certain stands of *Myrica faya* in the national park. The leaves turned yellow, branches drooped, whole trees died. The phenomenon seemed to bypass trees that had invaded the rain forest; it was far more prevalent in the arid sections of the park, where the dieback seemed to be rippling outward in concentric circles. At first, biologists suspected a disease, perhaps a bacterium that had made its way from one of the guava plantations on the island. In fact, the culprit was the two-spotted leafhopper, *Sophonia rufofascia*, a Chinese insect that was first noticed in Hawaii in 1987. Foote swept a hand net at the branches and caught one. It was a splinter of a bug—yellow, like one of those forgettable flecks that fly up if you beat some bushes with a stick. Two tiny spots at the back end marked the species. In large numbers, it can defoliate a tree as effectively as any herbicide. In the vicinity where we stood, Foote said, roughly sixty percent of the *Myrica* were failing. By and large it was a kingdom of the dead.

This would be good fortune—an invading tree felled by an introduced insect—if not for two concerns, Foote added. First, the range of the two-spotted leafhopper is restricted to the dry forests, whereas *Myrica faya* is found throughout the park. Second, and more troubling, the two-spotted leafhopper does not limit itself to *Myrica faya*. "We're starting to compile a list of rare and endangered species that are sensitive," Foote said. "What we're finding is that the native species likely to be vulnerable to the leafhopper are the ones that are most common. Dodonea is one of the most common shrubs in the park; it's very infested. O'hia is infested. About forty percent of the o'hia around here are dead. So there's no real evidence that *Myrica faya* is going to somehow be replaced by a native species. On the other hand, *Myrica* does seem to be disproportionately affected. There might be a long-term transition here to more of a shrubland rather than o'hia scrubland. Anyhow, right now Resources Management is sort of taking a wait-and-see attitude. There's really not any way to intervene."

In the interim, the park has assembled its own fleet of insects to combat *Myrica faya*. The first, a moth, *Caloptilia schinella*, was launched into battle several years ago by the U.S. Forest Service. Originally from the Azores, *Caloptilia* is a leaf roller: it starts at one end of the enemy leaf and rolls it into a tube, then lays eggs inside. This habit serves to protect the next generation of *Caloptilia*. More important—however unintended by the moth—this habit also takes the leaf out of the photosynthesis business; burdened with unproductive leaves, the affected tree soon dies. Unfortunately for biologists, however, *Caloptilia* is choosy: it assiduously avoids chlorotic leaves, the kind produced by the two-spotted leafhopper. Foote said, "Our concern is that even though the two-spotted leafhopper has achieved a good bit of damage to *Myrica*, our biocontrol agent may ultimately get knocked out here if the leafhopper continues to be a dominant force."

Foote got into the truck and headed back to park headquarters. Of course, he said, nobody expects one biocontrol insect to do the trick against *Myrica*; four or five, working in concert, will likely be required to reduce the tree's numbers. The leading candidates are various weevils that would attack the tree's roots and seeds. Ideal would be a moth that feeds on leaves: such moths are relatively easy to find, and they tend to be specific in their choice of plant hosts. Here, however, Hawaii's biocontrol effort is haunted by its past. Early in the century, in an effort to combat various moth pests then attacking major crops in Hawaii, agricultural biologists introduced large numbers of hymenopteran parasites: predatory wasps, different from the wasps occupying Foote's front lawn but with equally catholic tastes. These wasps made a sizable dent in the pest population; they also drove countless native moths and butterflies to near or complete extinction. More to the point, Foote said, any moth or butterfly released today to control *Myrica faya* or another invasive plant will be hit immediately by yesterday's biocontrol parasites. Even for *Caloptilia schinella*, the leaf-rolling moth that is showing some potential against *Myrica faya*, the days are numbered. "That's an irony I'm sure Frank Howarth can elaborate on."

By now the truck had reached the transition zone from windward to leeward. The day, once crackling hot and open, was thick again in mist and damp green forest. "It's a paradox," Foote was saying. "I have a lot of doubts about the efficacy of biocontrol in general, and I have some real concerns about its safety—particularly the biocontrol of insects, where there are numerous examples of major problems. As far as research is concerned, it's among the sloppiest I've ever seen. And yet from a conservation perspective in Hawaii, the biocontrol of weeds is really our last hope. There's no way that we're going to be able to deal mechanically with a species like *Myrica faya*. If we cannot get a successful biocontrol program going for *Myrica*, it will destroy the forest here. It's going to boil down to coming up with a much better program that is well funded. It doesn't cost much money. I mean, once you get an insect established, it's there and it is going to do its job and it's not going to take any more funding in order to keep it out there. That's the whole appeal of biocontrol: the long-term benefit is potentially permanent."

He gave a wry laugh. "But we're getting virtually nothing in the way of funding. Right now, with endangered species programs, we're just throwing money down the drain. Instead of focusing on habitat restoration, we're focusing on captive breeding of endangered honeycreepers that show no promise of long-term survival in Hawaii. I don't want to be too extreme here, but it is just . . . It makes me sick how much money is being put into a species like the Hawaiian crow. Any population biologist would tell you that there is no way that you're going to bring about recovery of that species."

Some weeks prior, the head of the U.S. Fish and Wildlife Service had publicly vowed not to let a single species go extinct on his watch. It was a noble sentiment, Foote said, and utterly irresponsible. "We're losing entire forests; we're losing entire ecosystems in Hawaii as a consequence of people who are unwilling to let go of a single species. By putting himself out on that limb, he is going to end up using huge amounts of resources that could benefit the long-term persistence of species that are common now but won't be if we spend all our money on the rare species. They're going to end up disappearing, and we're going to end up with trash forest: increasingly homogenous groups of species that have become introduced and established throughout the tropics. The only way to get around it is to get people to let go of single species. It will be interesting for schoolchildren to go to a zoo and see species that are functionally extinct in the wild and realize that they're still persisting in captivity. But that shouldn't be what our goal is. We should be looking more at trying to do something about the forests that we live in. It really rubs me the wrong way."

Foote reached the park headquarters. He pulled into an open parking spot on the grass, got out, and slammed the door. It was the first and really only time I had seen him so agitated. He trudged back to his office with a weary smile. "Well, that's my rant and rave. It really rubs me the wrong way."

In 1847, while Charles Darwin sat in his English laboratory examining dead Galápagos finches for the clues that would eventually point him to the idea of evolution by natural selection, a young naturalist in Massachusetts was conducting an experiment of his own. His name was Henry David Thoreau, and he had no interest in grand voyages. For him, no expeditions to explore the South Seas or to count the cats in Zanzibar, no search for the great hole that opened into what some believed was a hollow Earth. Instead he built a cabin in the woods and did his best to make an island of himself.

It was an experiment, by and large, in weeding. When not pondering the pond or walking into town to steal pies off windowsills (as legend has it), he spent his time in his garden "making invidious distinctions with the hoe," trying to make the ground speak beans instead of grass, and shaking his fist at the local woodchuck. He began with the notion that his garden would mark the boundary where nature ended and civilized bean society began, but he quickly came to understand that his was a boundary that nature would not respect. Particularly not the woodchuck, which left Thoreau conflicted: identify with it or eat it? (In the end he left it alone—a decision made somewhat easier by the removal, at age thirty-three, of all of his teeth.)

At issue was how to come to grips with nature's estranging otherness. Humans, one notices daily, are a bipedal paradox, simultaneously the products of nature and, with our big fat brains, the sole outside observers of it. "I may be either the driftwood in the stream, or Indra in the sky looking down on it," Thoreau mused. It's a weird evolutionary role to play, and thoughtful humans have never been entirely comfortable in it. Alienated from nature, we flock to it in such numbers that it is trampled in the process. Then again, if we define nature as everything alien to us, how can we feel anything but alienated?

Without quite realizing it, Elton confronted a similar divide. It was his notion, or at least his supposition, that a biological difference exists between "natural" invasions—the sort that have occurred through geological time—and those that occur today by the witting or unwitting hand of man. Undisturbed, evolutionarily mature, species-rich ecosystems are more resistant to the incursion of nonnative species than those simplified, "unnatural" ecosystems cobbled together by human activity, he contended. If that is the case—if nativeness somehow acts as a repulsive force—then native ecosystems must have a biogenic advantage over nonnative ones. Nativeness—naturalness—must matter in nature.

Elton hoped that invasion biology would account for this difference—and it has, all these years later, largely by dispelling it. "It seems clear that there is no prima facie case for the biotic resistance paradigm," Daniel Simberloff writes in an article in the journal *Biological Invasions*. The consensus among invasion scientists today is that, given the right opportunity, any native species can become an invader in some environment in the world; and any native ecosystem can be invaded by something. The most common factor determining whether an invasion succeeds lies not in biology—in some trait of the invader or the invaded region—

but in statistics, in the frequency and intensity of the introduction: in propagule pressure. Mark Williamson, in his book *Biological Invasions*, concludes: "It is possible to say something empirically about Elton's generalization, and that is that it is scarcely, if at all, supported by fuller data."

Of all things, Elton seems to have been led astray by semantics. Mark Davis, a biologist at Macalester College, has noted that in the 1960s, in the early days of invasion biology, most researchers used neutral terms like *introduced*, *nonnative*, and *founding populations* to describe the phenomenon. Elton was largely alone, though not for long, in his use of flashier terminology: *alien*, *exotic*, *invader*. While emphasizing the threat, the heavy use of this language over time has come to imply that the "otherness" of an invading species is somehow ingrained in its biological being. In true fact, an invader is simply a species that comes from elsewhere; its definition is purely geographical.

Our language troubles do not end there, Mark Williamson adds. There is a tendency to think of invaders, weeds, and colonizing species as a synonymous group, when in fact they comprise three separate and only rarely overlapping categories. An invader comes from elsewhere. A weed or pest is simply an organism we dislike; it may be native or an invader. (Giant ragweed is native to North America but a weed in the eyes of many North American farmers.) Weeds sprout from human opinion. Colonizing species are altogether different. They are ecology's nomads, impermanent to any one place, that persist by perpetually settling new ground and ceding it as the territory matures. Colonizers may be natives or invaders, weeds or not; of the three, only colonizers have something like a biological definition.

So to reliably predict that a given invader will or won't become a bothersome weed, simply by examining certain traits of the organism in question—as many agricultural companies and biocontrol scientists claim the ability to do—is an act of semiotic futility, Williamson writes. "The characters of invaders that become weeds might reasonably be expected to combine the characters of both invaders and weeds, but that is only possible if both invaders and weeds have definable characteristics." It is a search for scientific indicators where by definition no such indicators exist, rather like comparing an apple to an orange and asking when the former will become a vegetable. One invasion biologist I later met has disavowed entirely the term *invasive species*. Call them *invading species* or *invaders* if you must, he says, but *invasive* credits the organism with a biological advantage that science is not prepared to recognize. It may sometimes seem that how we talk about nature is irrelevant to how we deal with it, but with ecological invaders—nonindigenous species, if you will—it makes all the difference.

In any event, the concept of biotic resistance does not drive conservation planning at Hawai'i Volcanoes National Park. According to its official resource management plan, "Natural processes in ecosystems changed by alien species are not understood well enough at present to demonstrate that preserving native plant and animal communities will protect natural ecological and evolutionary processes." As Foote expressed it one afternoon, "We're fairly myopic here. We tend not to worry too much about whether a generalization can be made from a particular park research program. We're more concerned about doing a good job of restoration and trying to halt the invasions that are taking place."

We had driven over to the dry side of the park to take a look at a stretch of forest called Kipuka Pua'ulu, one of fifteen Special Ecological Areas that form the backbone of the park's management plan. They are test tubes in their own right, large and largely self-sustaining assemblages of native species. Lava flows from Kilauea Crater or smaller craters nearby; it runs downhill in streams and rivers, carving through the forest, flattening wide stretches of it, leaving other patches unscathed—green islands stranded on a black sea of hardened lava. So it is that the Big Island, and every island in the Hawaiian chain, is itself an archipelago, itself a necklace of islands, of pearls of habitat marooned one from the next. These islands, called kipukas, see some traffic between them: slender-billed honeycreepers traveling far afield to feed on blossoms and, incidentally, pollinating them; seeds of koa and o'hia and mamane drifting on the wind. But the tiniest, least mobile inhabitants of these kipukas—the drosophilas, the soil mites, the caterpillars, the snails—might as well be on the most remote Pacific island. They might as well be in Hawaii.

On these colonists, isolated in small communities, natural selection works with renewed vigor; the effects of inbreeding and random genetic drift are amplified. Often new species arise, such as flightless wasps and hammerhead fruit flies. And often such species are confined to the kipukas in which they arose. Eventually the lava flow that surrounds their kipuka shows sprigs of life; over decades and centuries it reverts to forest.

But even for a long time after, the integrity of the kipuka remains intact and distinct: the age, depth, and chemistry of the soil, and the assembly of mites and springtails and microscopic organisms that inhabit and revitalize the soil; the suite of trees and smaller plants that grow only on this kind of soil; the weevil that feeds only on these fronds; the damselfly that breeds only in this leaf axil; the caterpillar camouflaged to match only this cloud-forest liverwort. To the extent that these kipukas are relatively self-contained, easily demarcated, and often represent unique suites of plants and organisms, they make up the majority of the park's Special Ecological Areas.

Most of the forests surrounding Kilauea are relatively young: roughly 90 percent of the total acreage has been completely obliterated by lava in the past thousand years, the biotic clock reset to zero. Kipuka Pua'ulu, in contrast, is one of the oldest fragments of forest in the park, having sprouted from a thick blanket of ash cast off by Kilauea at least two thousand years ago. It is dominated by tall, stately koa trees, once a common feature of the Hawaiian landscape but now rare. What is now Kipuka Pua'ulu once covered a much larger area, Foote said, but successive eruptions from Mauna Loa had come down and covered the surrounding forest. In fact, we were standing on the border of two kipukas: the older one, and a younger one only about four hundred fifty years old. It all looked like trees to me, but Foote, attuned to the subtleties of biological diversity, could easily tell them apart.

"Most of the Hawaiian forests aren't that rich, species-wise," he said. "If you were to go out and put up an exclosure, you'd save a chunk of forest, but it wouldn't necessarily be the best chunk. What we're trying to do is prioritize and preserve areas that are richest. That accounts for a relatively small proportion of the forest. If we focus our attention in a few areas, we might be able to preserve the bulk of the diversity that's present. In Hawaii, we have the luxury that many of the species that evolved here evolved in fragmented forests, whereas on the mainland, many of the species require much larger tracts of forest. There's room for optimism that even a forty-hectare tract of forest will be self-sustaining, simply because many of the species that evolved here are used to colonizing small kipukas of forest and persisting in them. The lowlands—they're gonna take a lot of work to do anything that results in a predominantly native system. The kipukas up here, once you do the initial work, it won't take much to maintain them."

All told, Hawai'i Volcanoes National Park spends a million and a half dollars annually on the control and removal of alien species; fully half of that, about eight hundred thousand dollars, goes to the removal of introduced plants. This is a modest amount in the larger war against alien species. Florida—which, like Hawaii, boasts a semitropical climate and a diverse cache of endemic organisms—annually spends seven million dollars alone fighting hydrilla, an aquarium plant from Sri Lanka that was dumped into a canal in 1951 and has since emerged as a sort of underwater kudzu, choking seventy-five thousand acres of state waterways. Everglades National Park spends half a million dollars annually combating two fast-growing alien trees, Australian pine and melaleuca, that crowd out native plants and the creatures that depend on them. At the conclusion of the federal government's 1993 report, "Harmful Non-Indigenous Species in the United States," the authors suggest that "the metaphors that guide resource management are shifting from the self-sustaining wilderness to the managed garden. The world is being defined more in terms of the 'unnatural' rather than the 'natural.'" Some environmentalists might wonder whether one can truly call an ecosystem "natural" if it requires constant management and maintenance by people. I put the question to Foote: What's the difference between this kipuka and a zoo or a garden?

"It's a world apart from an arboretum or a zoo," he responded after a moment's reflection. "Here you have a functioning ecosystem that requires very little in the way of manipulation. We're not feeding the birds. We're not planting the plants. The insects that pollinate the plants serve as food for the birds and reproduce on their own. The only thing we do is keep out aliens."

One evening I was strolling with Foote around his small lawn, discussing the plight of Hawaii's biological diversity, when he stopped suddenly with an inspiration. "Hang on," he said, and went inside. He returned a moment later, the screen door banging shut behind him. In his palm he held a large gray moth: a specimen of *Eupithecia monticolens*, he said, that he had captured at his porch lamp the previous night. In caterpillar form, the animal thrives on the pollen of o'hia blossoms and the protein therein. That in itself is an unusual evolutionary adaptation, as most caterpillars eat only carbohydrate-laden foliage. At one time, Foote said, the moths of

Eupithecia monticolens blanketed the windows of Volcano at night, drawn to the lights inside. "Now you're lucky if you see one or two. Their decline seems directly related to the introduction of yellow jackets, which pluck them right off the o'hia blossoms. It's really dramatic to see." Unfortunately, he added, the yellow jackets are thereby reducing a key food source for forest birds. In its caterpillar form, *Eupithecia monticolens* is a primary prey item for three endangered bird species. There are perhaps eighteen species of *Eupithecia* caterpillar in Hawaii, all of them the evolutionary offshoots, it is thought, of *Eupithecia monticolens.* Some members of the genus *Eupithecia* are quite unusual, Foote said. He suggested that I stop by his lab the following morning and have a look.

When we met up, he handed me a glass vial. Stoppered inside was a slim leaf. On it, an inchworm sat erect. Its hind end was fastened to the stem by tiny legs; its segmented body sloped skyward like a miniature leaf-green escalator. This species is called *Eupithecia orichloris,* Foote explained. How long ago it evolved into its own species is unclear; the DNA work that might spell out its molecular phylogeny had not yet been done. Nonetheless *Eupithecia orichloris* has retained its ancestor's taste for protein. In fact, *Eupithecia orichloris* and its Hawaiian kin have taken this legacy to the extreme. They are the world's only known carnivorous inchworms.

Foote shooed several ordinary fruit flies from a holding vial into the proverbial lion's den. As they settled in, the inchworm stood twig-still. After a few minutes, a fly wandered into one of several sensitive hairs protruding from the inchworm's hind end. Instantly the inchworm struck, reaching over backward and nabbing its prey with—I watched now through a microscope—two sets of clawlike legs. These held the fly in lock while the mandibles of *Eupithecia* began to graze: first on the tail, then on the thorax, the wings, the still-wriggling legs, until all that was left after fifteen minutes were two amber, geodesic eyes. Then these too were punctured and devoured. *Eupithecia* paused a moment, then stiffened in anticipation of its next meal. Rather like the brown tree snake, the predatory caterpillars have adopted a surreptitious lifestyle. One is gray and spiny, resembling the decayed twigs on which it perches; another is ornamented with tubercles resembling moss. One stations itself along the edge of koa leaves; another chews out the midrib of hapu'u fronds and lies there in ambush. I fed more flies into the vial and was spellbound by the activity: *Eupithecia* lashing again and again, victim after victim reduced to a bolus dimly visible through its skin. I felt oddly sated, at last full of the marvel I had come to see.

Foote wandered in from across the hall with an additional bit of information. Like the native birds that feed on them, he said, these *Eupithecia* are inching toward oblivion. "Their prey are disappearing. Two species of picturewing drosophila that the caterpillars feed on are disappearing from the park."

The caterpillars were being simultaneously eaten and starved, pressed from both sides by the accumulating impacts of introduced species. For a moment it was as though the face of a watch had been pried off and the inner workings revealed to me: inchworms, flies, wasps, birds, mosquitoes, pigs, worms, trees, seeds—gearwork too intricate and interrelated for me to discern what time it was or why. Of course, that is just an analogy. Nature does not function precisely like clockwork, a tapestry, a cathedral, a pyramid, an airplane, or an international bank; metaphors are drawn from the world of human invention and knowledge, whereas nature is far larger than either of those things and has hardly begun to be understood. The biologist Robert O'Neill has proposed that even the concept of an "ecosystem" has outlived its descriptive usefulness. An ecosystem suggests a system: an integrated unit, self-regulating, resistant to disturbance, a collection of dynamics that occur within set boundaries. Whereas in fact, the order of the day is constant change. Species disperse and invade, come and go, evolve and go extinct. Any organism can be an invader somewhere. Every ecosystem—or whatever one calls it—can be invaded by something. This is true even in the absence of humankind. An ecosystem is stable over time not because its list of species remains forever the same, but because it varies—not in spite of disturbance, but because of it. Huffaker's microcosm of orange persisted only because it was persistently infused with new inhabitants. Stability is not an end result; it is a state that nature is forever falling into.

Biologists refer to this as the "nonequilibrium model" of nature, and it has largely supplanted the old balance-of-nature view of the world, at least among scientists. Its only real fault is its lack of sex appeal. Without the striking image of an unraveling tapestry or a broken watch or a crumbling edifice, how does a conservation-minded scientist impress upon the public the threat posed by a phenomenon like ecological invasion? If nature is always changing, how do you manage it? If a line can be drawn anywhere, how do we know when we've gone too far? "By admitting to some kinds of change, we may have opened a Pandora's box of problems," cautions the biologist Daniel Botkin. "Once we have acknowledged that some kinds of change are good, how can we argue against any alteration of the environment?"

At the start of the jet age, the anthropologist Claude Lévi-Strauss lamented that the variety of human cultures was even then dissolving into uniformity. "Civilization has ceased to be that delicate flower which was preserved and painstakingly cultivated in one or two sheltered areas of a soil rich in wild species," he wrote in 1955. "Mankind has opted for monoculture; it is in the process of creating a mass civilization, as beetroot is grown in the mass." The ecologist's lament is similar: As the human race spreads and its sphere of disruption widens, the natural world is winnowed only to those plants and creatures that can thrive in our wake. As nature's most pervasive invader, human civilization has become a force of natural selection in its own right. Owing to simple disregard, many of the organisms we see—and many more that we do not—are being selected against.

"We are increasingly dealing with a number of species that are associated with humans throughout the world," Foote said when I returned the caterpillar to him. "If that's what you want to live with—a small suite of a dozen, maybe two dozen species—then you can live with that, I suppose. You can argue that biodiversity has a utilitarian value. But it's an aesthetic issue for me."

It was a surprising sentiment to hear out loud from a biologist. For the longest time it has been fashionable to discuss nature in the third person, as an external entity against which the prudence of human activities can be objectively gauged. Elton pursued the study of invasion in part because in doing so, he imagined, he would uncover certain ecological rules that would guide and improve conservation efforts. *What would nature do?* Fifty years of subsequent research, however, have yet to reward that endeavor. Biological diversity is indeed valuable, but perhaps less for any advantages—stability, resistance—that accrue to nature and more for the advantages that accrue to us. "Probabilistic ecology does not suggest an ecological imperative," Botkin writes. "There are no balances to protect. Rather, nature is protected and promoted because we derive benefits from it, whether the benefit be aesthetic, spiritual, scientific, or economic."

One might describe it as ecology in the first-person plural—an ecology of a personal nature. It is a disarming notion, that the strongest argument for preserving biodiversity might rest on something so mercurial, so subjective, so intimate as a personal desire to live in a world that is biologically rich. Yet why else had I come so far? For scientists like David Foote, Linda Pratt, and Earl Campbell, homogenization presents a kind of domestic crisis: a dwindling sense that one's native environment—one's home, perforce one's being—is unique in the world. A traveler confronts the same threat turned inside-out. In a homogenized world, where does one seek out novelty, surprise, wonder? If everywhere looks the same, where is there left to go?

In the end, despite our best intentions, nature is not a reliable model for wilderness conservation. That is what I remember seeing most clearly through Foote's microscope that afternoon. Viewed up close, nature is heartless, mindless, raw, and insatiable; it is red in tooth and claw. However much we care about it or its more attractive artifacts, it does not care for us, nor even for itself. Humans enjoy the notion that unlike all other organisms, we stand one foot beyond nature, outside looking in. But we flatter ourselves. Nature is sufficiently fearsome that, in order to live and thrive, every plant and animal, consciously or not, must align itself against it. Nature is other, for everything in it. It is the force that through the green fuse blasts the roots of trees, chews the heads off tiny flies, drives our green age and perpetually destroys it. No, the problem is not that nature has ended. The problem is that nature cannot tell us where to stop.

14

There is nothing like an airplane flight to make one feel like a seed in a pod. Partly it is the orderliness of the affair, the tidy packaging. Partly it is the tending one receives, calibrated and carefully timed, as in a hydroponic garden. Mostly it is the physical proximity. Precisely wedged into a chair among chairs, the passenger cannot avoid sensing that seating arrangements were made by a computer algorithm that aims to balance two commercial equations: the maximum amount of passenger income that can be squeezed from a given area of floor space versus the maximum amount of discomfort that a passenger will pay to withstand. Biologists sometimes classify organisms into two categories based on reproductive strategy: r-selected species, such as dandelions, which effectively toss overwhelming numbers of seeds onto the wind; and K-selected species, which produce relatively few offspring but expend large amounts of energy ensuring their comfort and survival. On a passenger jet, the K-selected group is called First Class.

One could argue that the propagule sensation is inherent to the very idea of the journey. Travel, one is told, is a seminal experience. The advertisements, the four-color brochures, the narratives all extend a similar promise: you will see, you will absorb, you will gain, you will grow. In some small way, you will be transformed. Travel is a weekend away, a reward upon retirement, a chance gift won in a game show or a sweepstakes. *Honey, we're going to Hawaii!* Applied by biologists to nonhuman organisms, the phenomenon is known as the ecological sweepstakes, and it explains how life arrives at a place like Hawaii to begin with. In the 1960s, Linsley Gressitt, an entomologist at the Bishop Museum in Honolulu, flew a small plane above Oahu and towed a net through the air to gather and study the extent of the "aerial plankton" up there, much as Darwin did on the deck of the *Beagle.* Gressitt collected thousands of seeds and insects belonging to dozens of species. More recently another entomologist calculated that on a fine day in May, a volume of air one mile square extending from twenty feet above the ground to an altitude of five hundred feet contains thirty-two million floating spiders and other arthropods. "This amounts to 6 arthropods per 10 cubic yards of air," he notes. "Ten cubic yards is quite a small space, about the size of a small clothes closet." An airline passenger should be so lucky.

By and large the cause of biogeographical uniqueness—the reason why plants and animals first arrived where they did and not somewhere else—is not the biological richness or paucity of the host environment, nor some biologically superior aspect of the invader, nor even something as mundane and seemingly pertinent as climate. Rather, the deciding factor is propagule pressure: the number of seeds that show up and the frequency and persistence with which they do so. Nature runs a numbers game. This was the case even before the rise of humankind. In 1974 the biologist Sherwin Carlquist examined the suite of endemic plants currently living in Hawaii and then worked the taxonomy backward to calculate which and how many ancestral species would have been required to colonize the island and how they first arrived there. Some, he determined, floated in on ocean currents (14.3 percent), aboard floating logs or vegetation (8.5 percent), or adrift on the wind (1.4 percent). By far the most significant vector of plant introductions to Hawaii, however, was the travel of birds. Plants like the ohelo, an endemic berry, bear fleshy fruits that are often consumed by birds, which retain the seeds in their digestive tracts and later disperse them in their wandering. This mode of transportation probably accounted for a large proportion (38.9 percent) of the Hawaiian plant introductions over geological time, Carlquist concluded, as such plants are well represented today among the endemics. Other plants bear barbed or bristly seeds or fruits that probably arrived stuck to birds' feathers (12.8 percent), small seeds that readily adhere to the mud on a bird's foot or lodge in the crevices of its feathers (12.8 percent), or viscid seeds or fruits that simply stick to a bird (10.3 percent). Carlquist later conducted a similar study examining some hundred-odd herb species familiar to both North and South America. Virtually all of them possessed adaptations that would improve their odds of being transported by birds: barbed or bristly seeds or fruits (42.4 percent); fleshiness and edibility (19.9 percent); stickiness (18.9 percent); small size, ideal for mud or feathers (15.1 percent). For plants, at least in Hawaii's prehistoric days, the propagule pressure was applied chiefly by birds.

Increasingly, human flight offers a parallel byway, and not only for plants. The first official inspection of aircraft in the United States, in 1928, found ten species of insects hitchhiking in the *Graf Zeppelin*. Between 1937 and 1947 the U.S. Public Health Service inspected more than eighty thousand airplanes; a third of them were found to contain spiders or other arthropods. In 1975, the crew of an orbiting Apollo flight spotted "a super Florida mosquito" flying around in their space module; it was seen once and never again. In the past two decades, epidemiologists in Britain and France have reported numerous cases of "airport malaria"—incidents of malaria among urban residents who had not traveled to the tropics, but who live within an insect hop of an international airport. Malaria-bearing mosquitoes not only have arrived in London and Europe aboard incoming flights but also, in one case, managed to hitchhike out to the suburbs in the cars of airport employees. In another case, the malaria patient was the landlord of a public house in Sussex, several miles from Gatwick International; he is thought to have contracted the disease, through an intermediary mosquito, from one of the many flight attendants who regularly stayed there.

Once, I spent a day in the company of Todd Hardwick, who runs a flourishing business in south Florida catching stray animals—a roving menagerie that, because Miami is the nexus of the nation's pet trade, includes everything from stray cougars and emus to monitor lizards and Vietnamese potbellied pigs. Among his tougher cases was Tabitha, a housecat that escaped her owner during a 747 flight from New York to L.A. and disappeared into the bowels of the plane. She allegedly remained there for twelve days and thirty thousand additional miles, until negative publicity and the threat of a lawsuit saw the plane grounded and searched. Airline officials soon declared they had found Tabitha (or a cat sufficiently similar) alive, in a crawl space above the baggage compartment. Christa Carl, a New York psychic, subsequently claimed credit for helping Tabitha resolve a problem with one of her past lives and showing her how to emerge from the plane's drop ceiling. Hardwick, when I spoke with him, was skeptical. At the airline's invitation, he spent several hours probing the floors and walls of the plane with a snaking camera that is popular with plumbers, looking for Tabitha. All he found were the numerous small pressurizing vents on the plane's skin that open outward to the great blue sky. He quietly suggests that the real Tabitha never made it back to Earth, at least not safely. Curiosity: sometimes it introduces a cat; sometimes it kills it.

Evolution, at least on a remote island like Hawaii, is the titrated result of propagule pressure and genetic isolation—of flow, and the lack of it. For natural selection to affect a species, the species must first somehow arrive. Dispersal, or gene flow, must be frequent and regular enough to enable that possibility. If dispersal is too infrequent—if the propagule pressure is too weak—then an organism cannot become established. (Among other things, this explains why large mammals are rarely endemic to remote islands. It's hard enough for one individual to get there, but it takes two or more to establish a population, and the odds against that are steep.) On the other hand, if dispersal is too frequent—if so many of the same species arrive that the incoming gene flow becomes a continuous stream—then future generations will not differentiate from one another or from the ancestral stock. Only if the gene flow ceases at some point, or slows considerably, can natural selection get to work; only then will an island species begin, slowly over generations, to evolve along a different path from its mainland kin. Humans are ever more skilled at transcending geographical boundaries and crashing the gates of remoteness, yet we do occasionally create genetic islands with our wake. In recent years, entomologists have discovered a variety of mosquito unique to the London underground—three of them, in fact, one each along the Victoria, Bakerloo, and Central lines. Introduced to the tunnels when construction began in the nineteenth century, the mosquito has been pushed on cushions of air to the far reaches of the Tube system and left there to evolve, Morlocks of the genus *Culex*.

The question today is how to balance the conflicting need of an island to be simultaneously unplugged and plugged in. "How do we remain the hub of the Pacific," Alan Holt, the director of the Nature Conservancy of Hawaii, asked me rhetorically at one point, "yet also maintain the value that isolation brings?" The dilemma applies equally to human communities and to biological ones. Before leaving Hawaii for the mainland, I spent a few days on the island of Maui, where residents were locked in heated debate over the virtues and hazards of a proposed plan to lengthen the runways of the local airport. Longer runways would effectively internationalize the airport by permitting the takeoff and landing of larger planes from more distant places. No longer would passengers from Asia and the East Coast of the United States need to first stop in Honolulu to transfer flights—a two-hour addition to an already lengthy journey. Faster access would mean greater passenger flow and would boost tourism, the only significant legal economy on Maui. But more international flights would bring more international passengers of all species—more incoming alien seeds and insects borne on the wind that blows from the airport twenty miles uphill to Haleakala National Park, whose managers staunchly opposed runway expansion. For a while, until airport officials finally tabled the expansion project, pineapple growers, flower farmers, retired surfers, and native Hawaiian activists sat down with hotel operators, park biologists, airport engineers, and of course lawyers and discussed the future of their island at such depth and length that the various cultural and biological concerns began to grow together in a tangle and almost seemed synonymous.

"We are a tiny little island in the middle of the Pacific, completely dependent on planes bringing money into our economy. We all agree that some kind of effort is needed regarding the alien species introductions. We know from our satisfaction study that the environment is the number one reason why people come to Maui. But we can't do it at the expense of economic stability."

"When you have to spend an entire day coming or going, if you have a seven-day vacation, two of those are due to travel. Some people say, 'It doesn't matter how long it takes.' But it wears off. The paradise stuff wears off, and you start to think, How can I get there faster?"

"The island is a microcosm of the whole country. The airport encourages more business, creates more jobs. But more development means that more people move here. You end up in the same place you were, but now you've lost everything."

"We can't stop every single alien species from coming in. Even the scientists say so. Stuff that's on Oahu is eventually going to make its way here. Maybe it'll take five years; maybe it'll take ten years. But all we're talking about is a delay for that length of time."

"There are so many people here, they've changed the face of Maui. They're not in touch with what's native about it. Nobody is saying don't grow, don't develop. They just question the pace of development."

"In some ways, the views of tourists are more important than the views of locals. In some respects, it has gone too far. It's not what it used to be. Maybe we're killing the golden goose. Even the tourists are starting to see it."

"You can't recall a bug. It's a time bomb: once you turn it on, you can't turn it off."

"We are very 'controlled growth.' We know that we need to keep Maui 'Maui' to continue to attract visitors. We are not Honolulu. We do not want to be Honolulu. That is not the goal of the tourism industry on Maui. That's not why any of us live here. I wouldn't give you two cents to live in Honolulu."

"There's so much competition out there. We've all become far better and more frequent travelers than we were in the '80s. You gotta maintain, or next year Mexico is gonna look pretty appealing."

"If this project makes most of the people happy, that's great. If it makes everybody happy, fantastic! I can't think of a case where that has actually happened. Everybody can't be happy. Hawaii is paradise, and still not everybody likes it here."

As an egg or a seedpod matures, the propagule within grows and begins slowly to test the walls of its casing. So too aboard an airplane. As the in-flight hours pass, everyone and everything seems to swell. Papers spill from briefcases and book bags onto the neighboring seat; blankets, dispensed from overhead bins, wind up in a tangle underfoot. Even your neighbor's physical person seems to have swollen, encroaching now on your armrest; he seems incapable of letting a moment pass without some sighing, corrective adjustment. The air, stagnant and recycled, swells in your head. Propagule pressure. When the plane touches down and begins a halting process toward its gate, it will be all everyone can do not to burst from their chairs and scatter through the door, like the germs of jewelweed.

For the next few hours, however, there is only suspension, hibernation, diapause—an increasingly interminable wait. The plane hurtles through space, covering an expanse that not so long ago would have taken weeks to cross. You are going nowhere, fast. You take up a book. Through tinny headphones you absorb the musics of the world, which sound alike and eventually repeat. You peer out the porthole into a deepening blue. Is that the sky or the sea? Are you the driftwood in the stream or Indra above looking down on it? You sleep the earthen sleep of spores and dream their floating dreams.

Setting Sail

15

"Island biotas are very interesting to me," Jim Carlton was saying. "How organisms get where, the whole sweepstakes phenomenon that geographers have written about. Marine people don't think about those things very much."

James T. Carlton is a marine person. At the moment, he was at the helm of a rented minivan, navigating the twists and turns of the John Muir Parkway along the upper reaches of San Francisco Bay, north of San Pablo, not far from the town of Hercules. On board were three other marine people, a trio of marine invertebrate biologists Carlton has come to know over the course of his career. Another carload of marine people was somewhere out there on the highway, though whether they were in front or behind, Carlton couldn't say: his cell phone had stopped working. On the surrounding hillsides, under a bright June sun, plaid-panted golfers roamed acres of seamless green. Off and on for the past few years, around this time of year, Carlton has led a flotilla of colleagues on a biological survey of the Bay. For several days they scurry over docks, wharves, and tidal zones and gather representative samples of the local marine flora and fauna for later study: sponges torn from Styrofoam floats, gribbles and barnacles scraped from wooden pilings, whatever wriggling or microscopic creatures come up in the plankton net. These were all on board now too, in specimen jars in the rear of the van, adrift in dead seas of formalin.

Marine biology is the study of the sea, its saltwater margins and the life therein. Of the latter, the vertebrates—fish, whales, seals, and other bony animals—tend to dominate the public imagination. Lesser-known yet far more abundant are the invertebrates, the constituents of Carlton's expertise. There are barnacles, of course, and mussels and clams, oysters, shrimps, krill, crabs, lobsters. There are seaweeds, sea stars, sea squirts, and sea anemones; nudibranchs (sea slugs), pycnogonids (sea spiders), and holothuroids (sea cucumbers). There is the sea hare, a kind of snail that if roughly handled squirts purple ink, which it manufactures from the red algae it eats. There is the sea gooseberry, a small, roundish comb jelly that snares plankton with two sticky tentacles that extend from the sides of its body. There is a variety of sea squirt, a pinkish slab of marine flesh, known as sea pork. There is a sea slug scientifically named Doris. There are tube worms and ribbon worms, peanut worms and flatworms; syllids, spionids, cirratulids, capitellids, serpulids, spirorbids, and phyllodocids. There is the fearsome chaetognath, or arrowworm, which, seen under a microscope, I swear looks like a glass penis with teeth. There are medusae and Hydromedusae, Ctenophora, Chondrophora, isopods, amphipods, ostracods, and pelecypods. Barely more than cells are the diatoms, the dinoflagellates, the radiolaria and rotifers, the nannoplankton and picoplankton. On and on. Add to this list the myriad juvenile forms of marine organisms: the veligers of snails, the nauplii of crabs and shrimps, the seven instar stages of the larvae of barnacles, the planula of coral, the pentacula of the sea cucumber, the megalops of brachyurans, the scyphistoma of Scyphozoa. There is a flat jelly called *Velella* that, with a protuberance that rises from its body like a small sail, is blown along the surface of the sea. There is a sea strider called *Halobates*. There is a vast and barely charted world of marine plants and animals out there, a Peloponnesus of sea dwellers—most of which, quite frankly, until I began spending time with Carlton, I never knew existed.

The habitats of these organisms are equally—almost infinitely—diverse. Estuaries, tributaries, bays, shores; seafloor, rocky coast, mudflat, mid-ocean. Some organisms, the floating seaweeds, flourish only at the very surface of the water, along with their epiphytic attachments: snails and worms that live only here, like the orchids that grow solely on the upper branches of certain rain forest trees. Others—various shrimps, worms, and marine pill bugs—rise from the bottom mud into the water column only in the darkness of night. Much of marine life is concentrated in the photic layer, the top two hundred feet or so of water that, under ideal conditions, is suffused with sunlight. But the sea below is teeming too, feeding on "marine snow," the constant microscopic manna that precipitates from the sunlit layer down into the crepuscular realm: living and dead phytoplankton, exoskeletons freed of purpose, fecal pellets, detritus. Night feeds on day. The smallest plankton alight on mere particles and commence decay; nutrients are freed up; single-celled animals climb aboard. The tiniest ecosystems are no more than seaflakes falling into darkness. In short, although it does not appear so to the casual eye, the sea is a lattice of aquatic islands: a watery reticulate of subdivisions and sub-subdivisions, each defined by stern differences in salinity, temperature, and the availability of light and nutrients, each marine island as isolated from the next as the islands of any terrestrial archipelago. "The ocean is not just this homogenous slurry of water," Carlton said. "It is made up of distinct water masses whose biota sometimes don't overlap whatsoever. Take a boat from Boston and head for Europe. Sample the plankton around Boston Harbor. Two days out, you'll be in the North Atlantic Drift, and not a single species of plankton will be the same. The biota is completely different. The water is completely different: the temperature, the salinity, the amount of dissolved oxygen. The high seas, for an organism that lives on the coast—you might as well take an organism from the forest and throw it into the desert."

A few organisms transcend these boundaries and travel widely. Most of them are macroscopic, and most of them are swimmers: the oceanic salmon that returns to the freshwater of its native river to mate and spawn; the freshwater eel, whose life cycle takes it from the Sargasso Sea near Bermuda to the inland fens of England and back again. On rare occasions, a natural raft may carry palm trees or land snails or iguanas or other terrestrial organisms from one remote shore to another. The planktonic brood of the Caribbean spiny lobster tours the Atlantic basin: newly hatched nauplii ride the Gulf Stream north to Nova Scotia, catch the North Atlantic Current east to Europe, are swept south to Africa by the Canary Current, and arrive, twelve months later, back in the Caribbean as mature lobster larvae. But that is the anomaly. Most marine invertebrates—at least the ones small and light enough to be borne on a current, yet hardy enough to survive the transition from one marine zone to the next—are short-lived: their lives are typically measured in weeks or days, too brief to last the months required to drift across a wide sea.

"They'd have to go through more than one generation," Carlton says. "They'd have to reproduce along the way. That's like trying to get a redwood tree to cross the Sahara by having it drop seeds, grow, then drop seeds, until you had a thousand-mile-long line of trees." Some species of shipworm have been seen living in logs in mid-ocean, leading some scientists to mistakenly formulate what Carlton calls the shipworm syllogism: shipworms live in wood, wood floats, therefore shipworms can float across the sea on logs. Not true, Carlton says: the species of shipworm seen on the high seas are not the same species seen along the coast, nor vice versa. "When you sample the high seas, you don't see logs or drifting things with estuarine species on them. If you launched a piece of wood and it somehow got to the mid-ocean, by the time it reached the shore again, it'd be covered with organisms of the high seas." Most marine plants and animals are restricted by their physiology to a limited and local marine island: a South Sea atoll, an Oregon estuary, the mouth of the river that feeds the Oregon estuary, a frond of kelp, an intertidal rock. They are homebodies. They are victors of the ecological sweepstakes: somehow, long ago, they or their evolutionary ancestors reached wherever they are today and have managed to pretty much stay put since.

Suffice to say, then, it would be unusual to find a Japanese sea star in the coastal waters of Tasmania, or a comb jelly from Long Island in the Black Sea in Europe, or the European green crab—a shore crab native to the Atlantic Ocean—on the Pacific Coast of the United States. Yet Japanese sea stars can be found in the coastal waters of Tasmania; they appeared around 1990 and, voracious predators that they are, pose a threat to native and commercial shellfish. In the mid-1980s *Mnemiopsis leidyi*, a comb jellyfish from the American coast of the Atlantic, was inadvertently introduced to the Black Sea. It has become so abundant, and its appetite for plankton and fish larvae is so limitless, that the invader now constitutes 90 percent of the wet biomass in the Black Sea; the local fishing industry, tottering already, has collapsed. Since the 1960s, *Carcinus maenas*, a predatory crustacean more commonly known as the European green crab, has become established in coastal waters as far removed as Nova Scotia and South Africa, to the alarm of fishermen and marine conservationists. In 1989, Andy Cohen, a biologist with the San Francisco Estuary Institute, discovered the green crab in San Francisco Bay, where it feeds on the same shellfish preferred by native shorebirds. (According to a journal article by another biologist, the green crab also eats *Potamocorbula amurensis*—a homely clam inadvertently introduced from Asia in the 1980s and now the most abundant mollusk in the northern reaches of the Bay—"like pistachios.")

Cohen is, with Jim Carlton, a cofounder and coleader of the semi semiennial San Francisco Bay expedition. He sat now in the passenger seat of the van, looking at a road map and offering navigational advice. In

1995, after several years of fieldwork and literature reviews, Carlton and Cohen submitted a report to the U.S. Fish and Wildlife Service on the prevalence and effects of introduced species in San Francisco Bay. At the time, they counted two hundred twelve exotic species in the estuary—from Mediterranean mussels and Japanese clams to a weedy salt-marsh cordgrass from the Atlantic Coast—whose net impact on the region's ecology, they summarized, is "profound." And every year that Carlton, Cohen, and their marine friends return to update the tally, they find something new to add to the list. Carlton and Cohen have called San Francisco Bay "the most invaded aquatic ecosystem in North America." They acknowledge that their assessment may be premature: their 1995 study in fact was the first in-depth regional study of aquatic invasions ever conducted anywhere. Since then, inspired by Carlton's model, similar surveys have begun in estuaries from Chesapeake Bay to Pearl Harbor to Apra Harbor in Guam. So far, however, San Francisco Bay remains the comparative model. What Hawaii is to terrestrial invasion ecologists, San Francisco Bay is to their marine counterparts: the yardstick against which the integrity of estuarine ecosystems is now measured.

"And we tend to be conservative about what we call introduced," Carlton said. "Anywhere in the marine environment, we may seriously underestimate the number of invasions because we simply don't know about all the tiny stuff: the worms, the protozoans, the filamentous algae, the diatoms and dinoflagellates—there are literally hundreds of species. We say there are two hundred and twelve introduced species in San Francisco Bay. There could be four hundred or five hundred. There are huge groups of organisms whose status we don't even talk about."

Carlton was steering the van with one hand. The other was dipping into a large bag of potato chips, having already plumbed a bagful of miniature candy bars. Breakfast was a cupcake with bright blue frosting. For all intents and purposes, Carlton is the world's expert on marine ecological invasions. Accordingly, he is much in demand, and he travels constantly: to speak at nonindigenous aquatic-nuisance species conferences in Norway, Australia, Hawaii, South Africa, Fiji; to teach classes in Connecticut, Oregon, and Argentina; to testify before Senate subcommittees; almost never to vacation. He has become something of a global organism himself, a connoisseur of the red-eye. He had arrived in Berkeley from the East Coast late the night before and had been up much of the night prior finishing an article for a marine biology journal. From what his colleagues can gather, Carlton is a graduate of the Thomas Edison school of sleep: three or four hours of true sleep a night, with several allusions to naps—visible to the observer as a momentary shutting of his eyes and nodding of his head—during the day. His schedule is that of a brain surgeon who has been on call for ten years running. Yet at all hours Carlton remains affable and enthusiastic. "He's cheerful," an amazed colleague said while loading up the van at seven o'clock that morning. "He doesn't get all grumpy and weird." Carlton is more or less trim, and graying hair graces his head in thick curls. Also, although he takes no notice of the fact, nor appears in any way affected by it, Carlton is notably diminutive: five feet three inches of barely contained energy. He is a mitochondrion of a man, powered largely, as far as I could tell, by processed sugar.

Like ecologists of terrestrial invasions, Carlton is less concerned with island ecosystems per se than with the diminishing distance between them. At one time the movement of species was governed strictly by physical boundaries: an isthmus rose between North and South America, and the species of two continents mingled. The same isthmus rose, and a sea of organisms was permanently parted, east from west, Atlantic and Pacific. Oceans divided continents; continents divided oceans. Coastal estuaries were rendered remote by the high seas between them; islands were isolated by unbridgeable gaps. Carlton said, "The whole understanding of the evolution of biota that we teach in biology is one of allopatric speciation: oceans and continents are barriers to each other. And it doesn't matter anymore. At least for the kinds of organisms and systems I deal with, none of these invasions would occur in the absence of humans. Nature simply does not exchange the marine biotas of western Europe and eastern Australia. It just doesn't happen."

For the terrestrial ecologist, history offers a modicum of solace, or at least perspective. The scientific exploration of life on land has been under way for more than four centuries. For four hundred years, since the first major expeditions were launched to navigate and chart the four corners of the earth, scientists have been taking careful notes: botanists, bug nuts, bird-watchers, zoologists, fossil hunters; collecting, describing, depicting, comparing, cataloging, storing away, keeping track. The natural landscape that these first scientists took note of was more or less, with some notable exceptions, the natural landscape as it existed prior to the first wave of European travel. Looking back, the contemporary ecologist is granted a relatively unobstructed view of premodern biogeography: how many of which terrestrial species were where in the world—before Captain Cook's pigs and the age of air travel and express mail—back in the prelapsarian era when physical geography reigned supreme. They have a baseline, shaky as that line may be.

Carlton is having a harder time of it. Marine biology, he points out, is a young science, barely two centuries old. The first recorded deep-sea sample—a bucket of seafloor mud—was collected in 1818. The United States Exploring Expedition of 1838 scouted whales, charted much of the Antarctic coast, climbed Mauna Loa in Hawaii, and confirmed beyond dispute that Earth in fact is not hollow, nor can it be entered through enormous holes at its poles. Plankton were scientifically unknown before 1847. The first expedition mounted exclusively to study the ocean, the voyage of the HMS *Challenger*, took place from 1872 to 1876; it discovered that marine life exists below one mile down, and using trawl nets, it retrieved forty-seven hundred previously undiscovered species. The first reference book on marine science, *The Oceans*, was published in 1942.

And now Carlton, after arduous years in the field and in the library, has come to realize that by the time marine biology got going, by the time scientists began taking careful note of which marine organisms were out there and where they were, those organisms had already been subjected to centuries of human-mediated reshuffling—"to a good deal of chess play," as Charles Elton phrased it in *The Ecology of Invasions by Animals and Plants*. For the terrestrial ecologist, history is a limpid pool. Carlton sees only a turbid swamp—"the murky antiquity," he calls it. Until Carlton began to plumb it, the depth of this unnatural history remained largely invisible to his contemporaries. If some marine organisms were cosmopolitan, if two members of the same species were found at opposite ends of the world, they were presumed to have been so since time immemorial: marine physics had made them so. Human travel didn't enter the equation, or so it was thought. So it still is, in the opinion of some holdfast marine biologists. To some, it has not yet dawned that the biological seascape encountered a century ago was not a seascape untraversed.

"Marine people don't think about those things very much," Carlton was saying in the van, near the limits of Hercules. In recent years he has been poring over a newly published series of scholarly volumes on the native marine invertebrates of the Pacific Islands. Among his own many projects, Carlton, with Lucius Eldredge of the Bishop Museum in Honolulu, is writing a monograph of the introduced marine invertebrates of the Hawaiian Islands, a catalog that now stands at some three hundred species. What fascinates him is the number of species in the published volumes—the catalog of organisms that presumably count Hawaii as part of their native range—that would seem to be strong candidates for his own list of introduced organisms. For example, he finds it odd that several species of marine snail that are listed as occurring naturally both in Hawaii and in other parts of the world like Japan, Indonesia, and California, have larvae that exist as free-floating plankton for no more than three or four days. That is to say, a newly hatched snail larva would have only three or four days in which to drift from its eggshell, wherever it is, and to settle as an adult in whatever patch of sea mud or rocky shore it will occupy for the rest of its life. Yet Japan, Indonesia, California, even the Northwest Hawaiian Islands—the closest landmass to the Hawaiian archipelago proper—all are weeks or months away from Hawaii by drift current. If the snails are not strictly Hawaiian, if they are found elsewhere in the world, how exactly did they cross the ocean to get to Hawaii? "They ain't getting to Hawaii by drifting there," Carlton said. "Yet nobody seems to talk about that." He is aware that, once in a great while, even a snail can win the sweepstakes and make its way from one end of the world to the other. But this does not occur anywhere near frequently enough to create a steady gene flow, at least not without humankind's help. "A lot of biologists casually invoke the ocean currents—that the reason a lot of marine species are everywhere is because they've been carried around the world by ocean currents. Ocean currents don't do that. Ocean currents have not homogenized the ocean."

At heart, Carlton's endeavor is as retrospective as it is inspective. Other invasion ecologists divide the world of organisms into two categories, native and introduced. Carlton has uncovered a third category, an array of organisms that are neither clearly native nor alien; their true geographic origins are lost in the shuffled prehistory of marine biology. Their pasts are indeterminate, perhaps indeterminable. Carlton calls these species "cryptogenic." Their very nature precludes an exact estimate of their numbers; Carlton figures the tally is large, at least a thousand. In San Francisco Bay alone, he and Cohen have found at least one hundred twenty-three species that they consider cryptogenic. Status unknown. The actual number, he says, might be twice that. In effect, Carlton has stumbled upon an epistemological frontier. Most explorers discover something tangible: a sea, a continent, a Pacific archipelago, a lost civilization. Carlton has found the opposite: a gaping pit where solid ground was thought to be, a hole in common knowledge. In one scientific paper he asks, "What were the coastal oceans like in 1899, in 1799, in 1699? No one knows: it's embarrassing to say that we lack even a rudimentary synthetic picture that would provide the first answers to these questions." Carlton has no interest in making history. He just wants to find it.

In 1984 Carlton and a group of college-age students dropped five hundred glass bottles, corked and sealed with wax, into the waters off Cape Cod, to see where the sea would take them. Each bottle contained a slip of paper with a message written in five languages:

> HELLO! This drift bottle is part of a long-term scientific research project, on the movement of ocean currents, being conducted by students of the Williams College–Mystic Seaport Program in American maritime studies.

The message noted the date and location of each bottle's release and urged its discoverer to reply and indicate when and where the bottle had been found. Carlton and his brood repeated similar exercises through the 1980s, adding several hundred more bottles to the worldwide flow. Mostly they waited—days, weeks, often months, sometimes years. Responses trickled in from farther afield with the passage of time. He heard from third-graders in New Jersey and Portuguese sailors in the Bahamas. A beachcomber in Rhode Island advised Carlton to improve the watertightness of the seal. A man in the Azores asked Carlton to sponsor his immigration to the United States. A German respondent expressed delight in having discovered a message in a bottle on a beach in France: "For the first time in my life I had such an experience and it is for me a little adventure."

The replies all arrived at the same address, that of a small stucco building across the street from the Mystic Seaport in Mystic, Connecticut, at the administrative offices of the Williams College–Mystic Seaport Maritime Studies Program. Carlton, among his many professional responsibilities, is the director of the program, which offers otherwise land-locked undergraduates the opportunity to spend an academic semester in Mystic studying the literature, history, politics, and biology of the sea. Mystic in the mid-nineteenth century was one of several bustling, polyglot seaports along the New England coast. Men set sail for Calcutta, Hong Kong, Gibraltar; affluent women wore Smyrna silks and Turkish satins to church; the days sang with green parrots and Java sparrows brought from afar. Seamen and their families resettled from Mongolia, Africa, Cape Verde, Hawaii. Any Queequeg would have felt at home, or close enough to it. Today the seaport at Mystic is the Mystic Seaport, a cluster of quaint shops and a museum set on a small harbor. The decks of refurbished sailing ships crawl with tourists from Germany, Japan, Iowa. Once in a while, weather permitting, one of Carlton's students stands on the dock and provides passersby with a short seminar on invading marine species. The students live a block away in boardinghouses. During the course of their academic study, they will spend ten days doing fieldwork on the West Coast and two weeks sailing on the high seas. They must learn a maritime skill such as knot tying or celestial navigation. There are no term papers, only research papers. "This isn't 'Put on a lab coat and play at science,'" Carlton says. "This is 'Get out and find out how hard it is to extract information from nature.'"

Carlton's office—the neatest one of several he occupies, the one in which he greets visitors—is upstairs. It is a repository for an eye-opening collection of decorative crabs: plastic windup crabs, clacking wooden crab toys, crab-shaped dog biscuits, a crab-shaped catnip sack from Vermont, an oversize crab-shaped pillow he bought for fifty-three dollars at the Baltimore airport, and, across the room, a crab lantern, its door open, into which Carlton shoots rubber bands when sitting at his desk. At the moment, however, he was seated downstairs in the visitor waiting room. A globe of the Earth sat on a coffee table in front of him, and he used it to point out the world's oceanic currents: the Gulf Stream, the Southern Equatorial, the North Atlantic Drift. More than anything, Carlton has devoted his career to the study of biological flow—the motion of organisms of all kinds, by all means, to and from all places, from ancient to modern times. "Bioflow. How do things get where? What are its limits? I've got files on all possible natural vectors. Aerial dispersal; things on birds' feet, insect wings. I want to develop a thorough background." He was waving his hands over the globe like a magician. "Tens of millions of years of bioflow. Suddenly out of Africa comes this primate, and within ten thousand years it's an organized vector. Within the last five thousand years it has unraveled all the borders. Ours has become a borderless world."

The bounds of Carlton's knowledge, such as they are, are manifest in a white clapboard house up the street from his office, where he keeps his personal library. I once spent several days there at his invitation. The house had several rooms, every inch of which was awash in words. Bookcases stood floor to ceiling everywhere: a shelf for books about the Pacific Coast (*A Quantitative Study of Benthic Infauna in Puget Sound*); a shelf for books about mollusks ("my specialty," Carlton said proudly); numerous shelves packed with maritime history (*Voyage*; *The Seacraft of Prehistory*; *Sea Routes to Polynesia*). Field guides, ecology textbooks, monographs; *Introduction of Foreign Species, Dispersal and Migration*; *Green Cargos*; *The Giant African Snail*; *Immigrant Killers*; *Animal Killers*; *The Alien Animals*. The collected works of Thor Heyerdahl; collectors' editions of John Steinbeck's *Cannery Row*. What had no shelf sat on the floor, on a windowsill, or on some other long-buried surface, which in turn served as a platform for any number of manila folders with scientific papers about one introduced species or another: Japanese salmon in Chile, Pacific clams in Ireland, Atlantic striped bass in California. One shelf held an assortment of marine biology serials and journals: *Oceans*, *Oceanus*, *Pacific Science*, *Amphipacifica*, *American Conchologist*, *American Malacological Bulletin*, *Hawaiian Shell News*, *The Veliger*. Another shelf, which continued down a long hallway, held Carlton's collection of early books about evolution, including an 1868 edition of Ernst Haeckel's *The History of Creation* and Charles Lyell's *Principles of Geology* from 1830. This hallway was made considerably narrower by the occupancy of several filing cabinets devoted, if the labels were any indication, to specific introduced species: INTROS—MARSH PLANTS or INTROS—BARNACLES or INTROS—PANAMA CANAL, SUEZ CANAL, SALTON SEA.

I counted almost two dozen four-drawer filing cabinets in all, each one stuffed to bursting and piled high with more files and papers, including one entitled "Faecal Pellets." Sitting there, struggling to ingest some fraction of this literature, I felt quite small, like an intruder in the burrow of some very industrious bookworm that was itself progressing through a sea of measureless proportion. Sooner or later my tunneling led me down a hall of shelves and files and papers to the lavatory. The bathtub held several cardboard boxes, each one full of specimen jars of marine plants and creatures suspended and preserved in a yellowish scientific liquid—samples from the San Francisco Bay expedition, Carlton said, which would soon be making a cross-country journey of their own, to proper storage at the California Academy of Sciences.

There are islands, and there is getting to them. There is the sea and the crossing of it. There are vessels: rafts, dugouts, skiffs, canoes, longboats, ferries, schooners, clippers, trawlers, yachts, tugs, barges and barge carriers, bulkers, freighters, colliers, tankers (carrying oil, chemicals, fruit juice), aircraft carriers, passenger liners, landing craft, destroyers. Some ships carry nothing but what are known as "containerized goods": containers for containers and the things they contain. And what is a container but a kind of island, a division of contents unto itself? In 1778 Hawaiians jostled on the sands of turquoise bays as Captain Cook's ships, *Resolution* and *Discovery*, appeared offshore. They nudged, pointed, speculated. Look, they said: floating islands.

To Carlton, the notion that ships are floating islands is not merely figuratively true; it is literally so, and his career—indeed, an entire branch of ecology—has grown up around its demonstration. Consider the wooden ship *Arbella*, the subject of one of the hundred-plus papers Carlton has written about maritime biology. The *Arbella* sailed from England on April 8, 1630; sixty-five days later it anchored off Cape Ann, Massachusetts. Its cargo, one hundred English settlers bound for the New World, was typical of the era. What interests Carlton is not the human colonists on board, nor the livestock, plants, and seeds they carried with them. Rather, Carlton's interest is the marine passengers, the fouling community of organisms that settled onto and into the ship's hull and traveled with and plagued virtually every ship until the modern era.

The *Arbella* almost certainly carried shipworms. The shipworm is a worm in name only; in fact it is a bivalve mollusk, like a clam, mussel, scallop, or oyster, albeit a naked one. Its body is soft and elongated and entirely unprotected by its two shells, which are exceptionally small and located at what one might call the creature's front end. The sides of the shells are lined with rows of fine teeth; viewed close up and head-on, it brings to mind one of those giant rock-munching machines that some years back chewed a tunnel beneath the English Channel. Although most shipworms grow no more than twelve inches long, some grow to three feet or more—the size of a brown tree snake, at least the ones in Australia. Other mollusks burrow into sand. The shipworm burrows into wood: it settles as a larva, files away at the hull with its rasped shells, and, as it grows, mines out a tubular cavern for itself, which it lines with calcium carbonate it exudes from its body. Two small siphons, located at the back end of the shipworm, protrude slightly from the burrow into the water, filtering and ingesting plankton. When the mollusk dies, it leaves behind its tunnel, a convoluted hollow with an eggshell lining. One burrow may intersect others, or the lot of them might collapse and form a vast hollow within the hull that is entirely invisible to the outside viewer. Carlton has inserted his arm up to the elbow in the collapsed caverns of shipworms. Like the larvae of various other bivalve mollusks, the naval shipworm, *Teredo navalis,* also has the unusual ability to survive passage through the intestines of creatures that chance to ingest them. A scientist in 1938 noted: "Since *Teredo* and other pelecypod larvae are able to withstand trips through the alimentary tracts of other animals, they may be conveyed long distances from their place of origin."

Presumably the shipworm got its start long ago on submerged stumps and sunken logs, then moved, as civilizations grew more ambitious, to dugouts, rowboats, schooners, dock pilings. If ever an organism was built to travel, the shipworm is it. Pliny and Ovid wrote about shipworms. Twelfth-century vessels in the Mediterranean carried shipworms. Christopher Columbus lost two ships to shipworms on his fourth voyage in the West Indies. As early as 1590, the ports of Brazil were so notorious for shipworm activity that English ships were forbidden to enter them. Even today a fisherman in the tropics may watch aghast as his small wooden boat, seaworthy one moment, sinks at the dock. The naval shipworm first appeared in San Francisco Bay in 1913. It became so abundant so quickly that between 1919 and 1921 virtually every major wharf, ferry slip, and pier in the northern part of the Bay collapsed from it, at a modern-day cost of three billion dollars. In 1993, workers in Manhattan began the urgent repair of several piers along the Hudson River that were newly in danger of collapse: a cleaner river benefits shipworms too. Tropical species of shipworm in turn have been found in Long Island Sound just a few miles from Mystic, thriving in the warm effluent of a nuclear power plant in Waterford. I spent many childhood summers on the beach at Waterford; I learned to sail there. I find it strange now to think that for all biological intents and purposes, I was charting equatorial waters.

So Carlton assumes that the *Arbella,* a typical ship of its time, carried shipworms acquired in one port of call or another. As the ship sat at anchor, other organisms also would have found their way aboard: algae and barnacles; clams and worms that rose from the harbor floor and settled into muddy cracks in the keel; crabs and even fish hidden in the tranquil, abandoned caverns of shipworms; and of course gribbles, a kind of marine pill bug that burrows into the hull and slowly chews away layer after layer of wood. Scientists speak of the "primary film"—an initial assortment of fungi, algae, and single-celled plants that settle on a hull or any submerged surface, as eager seeds might colonize a forest clearing or a fresh lava field. The marine colonists are bound fast by mucilage, a sugary glue secreted by associated bacteria. Upon this bed an entire ecosystem blooms, a luxuriant subaqueous jungle: seaweeds and sea anemones, sponges, sea squirts, sea slugs, flatworms and roundworms, limpets and other snails, and whichever crustaceans grab hold as the ship begins to move. Carlton can only guess at precisely which organisms were aboard the *Arbella.* Nonetheless, he writes, "there is no question that the *Arbella* was a floating biological island." The fouling community on a typical wooden ship might be a foot and a half thick and would slow the craft considerably. Some vessels of the period covered their hulls with layers of tar, animal hair, and "sacrificial" elm (later, cast sheet leading came into use) as prophylaxis against fouling. But as many vessels or more used no protection whatsoever.

Carlton pictures the *Arbella* setting forth, plowing into the high seas, its next shore two months away. How many colonists survive? Which propagules shall propagate? The casual hangers-on, lackadaisical crabs and fishes, fall away early on. Others die en route, shocked by drastic changes in temperature and salinity, starved for plankton in mid-ocean. Yet others remain firm, ensconced in their tunnels or shielded within tentacles of seaweed or sleeping the hungerless sleep of spores. The new port is another estuary, the marine climate perhaps not so different from the port of departure. The *Arbella* anchors, lingers long enough for some of the organisms to feed, grow, and reproduce. Barnacles send forth young nauplii; anemones produce planulae; translucent medusae, nascent jellyfish, break off from their parental polyps like flower heads drifting from their stems. Over time the port becomes a kind of aquatic transit lounge, with sea creatures coming and going, taking residence or not. I picture frequent fliers at passenger gate, travelers so familiar with the anonymity of waiting areas it is as though they live there, or the transient men and women one sees on the street late at night.

In 1987, to better understand fouling organisms and how ships of old may have dispersed them, Carlton and a colleague, Janet Hodder, spent two months following the travels of the *Golden Hinde II*, a replica of the three-masted English ship that circumnavigated the globe under Sir Francis Drake from 1577 to 1580. The replica, built in England in 1973, had made its way to Yaquina Bay in Oregon; Carlton and Hodder kept up with it as it traveled slowly down the coast to San Francisco Bay. The ship's hull had been treated with copper-based antifouling paints; Carlton nonetheless found thriving communities of organisms on the hull, keel, and rudder. In addition, for experimental purposes, they attached panels to the rudder on which various fouling organisms soon settled. In each port of call along the way, they removed the panels, examined and tallied the survivors of ocean transit, and attached new panels in their place. In this manner, Carlton compiled a detailed measure of which organisms survived the various stages of the journey, and the temperatures and salinities they weathered along the way.

Over the course of the voyage Carlton and Hodder counted sixty-four major taxa, or groups, of species, including fifteen kinds of mollusks, twenty-seven kinds of crustaceans (a group that spans everything from crabs and gribbles to water fleas and fish lice), and seven kinds of polychaete worms. Although not all the organisms were found on every stage of the journey, the survival rate from port to port was notable. Of the twenty-two taxa that boarded the *Golden Hinde II* in Yaquina Bay, twenty-one were found when the ship arrived a day later in Coos Bay, Oregon. Of the fifty taxa that left Humboldt Bay in California, forty-six—92 percent—survived the three-day voyage to San Francisco Bay, among them young barnacles and sea squirts that evidently were born en route. A notable casualty was *Dendronotus frondosus*, a carnivorous nudibranch, or sea slug. Nudibranchs of this genus are highly mobile, wandering freely through a marine forest, grazing on hydroids. The slug breathes with the aid of elongated fleshy gills that rise from its back like branches; the animal resembles a small thornbush. Carlton found that *Dendronotus frondosus* did not survive the first leg of the ship's journey from Yaquina Bay to Humboldt; however, its cousin *Onchidoris bilamellata*, a slower-moving slug with the low profile of a limpet, was more steadfast. Carlton surmises that the larger, faster-moving slugs are more likely to be washed off during an ocean passage, victims of the old bleacher adage "Move your feet, lose your seat." Nevertheless, over the decades and centuries, large sea slugs have dispersed far and wide around the world which suggests to Carlton that they traveled under the protective canopies of what must have been massive fouling communities.

And all of this, Carlton writes, speaks only to what was carried on the outside of a sailing ship. To stay trim and stable in the water when sailing empty, cargo vessels filled their holds with dry ballast: stones, sand, gravel, or whatever debris was on hand. Upon arrival, goods replaced the ballast, which was heaved into the channel or, when that method was outlawed, piled ashore. Thus the solid earth of nations made its way across the seas. Ballast lots sprouted in the major cities of the U.S. East Coast; local botanical clubs soon flocked to view the exotic flora—bristly oxtongue (*Helminthia echioides*) from England, black bindweed (*Polygonum convolvulus*) from Europe, both now common worldwide—that arose there. No world was as new as the shipyards of Baltimore and Philadelphia and Camden, New Jersey. "As I review these ballast deposits, and detect so many strangers," one amateur botanist noted in 1876, "I feel a reawakening of that interest which a ramble about our fields and woodlands fails to create." Sightings of new plant species filled the pages of botanical gazettes: one hundred twenty-five new species in Philadelphia; sixty-four in Pensacola; two hundred fifty-eight in New York. Natural history journals sprouted lists of "ballast plants." In 1879 another amateur botanist in New York reflected that the large balance of trade in favor of the United States "has compelled a great many vessels, for want of freights on their westward trips from Europe, to come more or less laden with ballast. At the Atlantic Docks, Brooklyn, and on Gowanus Creek, vessels have for many months past been discharging it without cessation, night and day." The economy literally blossomed. Manhattan streets were built on loads of ballast several feet deep: 107th Street from Third to Fifth Avenue, 100th Street east of Second Avenue. The weight of one mode of travel provided firmament for the next. Peruvian heliotropes bloomed on Eighth Avenue, European heliotropes in the Bronx. Although many of these "waifs from abroad" would perish after a few seasons, one botanist wrote, "sufficient opportunity will nevertheless be afforded to some, not hitherto reported here, to test their endurance of our climate and to compete with our native growths." The adaptive strategies pursued by the city's indoor fauna—the social climbers, the human wallflowers—were as well suited to the flora just outside the window. Be pushy or blend with the woodwork. Fit in or get lost. "The less hardy plants will be ejected by our vigorous weed."

By the 1880s the nature of shipping—and the nature inadvertently shipped—had begun to change. Iron ships came into regular use: metal-hulled motor-driven behemoths with cargo holds partitioned by iron bulkheads. The ships of today carry that trend to mind-boggling proportions: the largest ship currently afloat is the *Jahre Viking*, an oil tanker some fifteen hundred feet long. The vast communities of fouling organisms that once blanketed the hulls of sailing ships are largely absent from the modern cargo vessel. Moving through the water at twenty to twenty-five knots, such a ship is stripped free of many of the seaweeds, worms, sea slugs, and crustaceans that might otherwise take hold. (The *Golden Hinde* of the 1590s achieved an average speed of five knots.) Where a sailing ship of old might linger in port for weeks or months, the modern ship typically departs within a matter of hours, insufficient time for organisms to settle on board and assemble in any quantity. Fouling communities are a costly drag: a layer of algae only a millimeter thick can reduce a ship's speed by 15 percent, a coat of barnacles by nearly half. Consequently, the hulls of twentieth-century ships have been coated with successive generations of toxic antifouling paints, each one more effective than the last. (Some also damaged the environment, and have been banned.) The paints have reduced, though not eliminated, the role of the ship's exterior in the global transport of organisms. The Asian green seaweed *Codium fragile tomentosoides*, common now from Nova Scotia to the mid-Atlantic (and sometimes known as dead-man's-fingers or Sputnik weed), first arrived in New York as a fouling organism, Carlton believes. He also thinks about mobile dry docks and oil rigs, which accumulate astonishing amounts of biomass during their long tenures at sea, and "sea chests," those protected portions inside ship hulls into which water is drawn to aid propulsion, which typically go untouched by antifouling paints. And like malarial mosquitoes, some species of fouling seaweed have developed resistance to the toxins invented to eliminate them. Shipworms, at least, are little menace to ships any longer.

The most significant change in the shipping industry, from the point of view of a marine biologist, was in ballast technology. Ever-larger ships required ever more stones, dirt, and debris for stability; it became easier, quicker, and less costly instead to pump seawater into and out of empty cargo holds and segregated ballast tanks as needed. A container ship traveling empty from Japan to Oregon to pick up a load of wood chips might begin its journey by filling its ballast tanks and a central cargo tank with harbor water for stability along the way. The water will be released when the ship arrives in Oregon, along with whatever marine life that water contains. In 1989 Carlton collected a few jars of ballast water from such a ship arriving from Japan and took them back to the lab for inspection. His findings were seminal, in every sense of the word. In that small volume of water Carlton found more than fifty different species of marine organisms, none of them native to Oregon: crabs, shrimps, mussels, clams, diatoms, dinoflagellates, barnacles; predators and prey; larvae and adults; zooplankton and phytoplankton—representing every stage of life, every habitat in the water column, every level in the food chain.

In *The Ecology of Invasions by Animals and Plants,* Charles Elton wrote, "Accidental carriage in or on shipping, that is in water ballast tanks or on the hull, has been a powerful and steady agency dispersing marine plants and animals around the world." Elton provided a handful of anecdotes, cobbled together from the few scientific papers that mentioned the phenomenon. That was 1958. One or two scientific papers came and went—they're all on file in Carlton's office—but until Carlton came along, nobody had studied the phenomenon in any long-term, systematic fashion. When he started out, few of his colleagues and advisers even knew what ballast water was. As he says, "I might as well have been studying those plastic things at the ends of shoelaces."

By 2000, several dozen scientific papers had been written about ballast water, a notable fraction of them with Carlton's name attached. A retrospective look at ship-trade patterns and the biology and life cycles of certain marine organisms led Carlton and other scientists to conclude that ballast water has been a vector of invasions since at least 1910, when the Chinese mitten crab—notorious for burrowing holes in dikes and levees—suddenly appeared in rivers in Germany. (It has since spread to San Francisco Bay, among other locations.) Carlton believes that the rate of ballast-water introductions has increased in the past two decades, and not simply because more researchers are investigating the phenomenon. The rate of global trade has doubled every seven years for the past several decades; some 80 percent of that trade is conducted by ship. On any given day, thirty-five thousand commercial and private ships are in motion, carrying—and dumping—billions of gallons of ballast water: from Seoul to San Francisco, Tokyo to Tasmania, the Black Sea to the Great Lakes, Brazil to Seoul.

Carlton, in effect, is a biologist of human commerce. Other ecologists conduct their fieldwork in rain forests, coral reefs, or tropical atolls, the kinds of settings that urbanites work and save for years in order to visit just once. Carlton's field sites are about as far from the classical definition of a natural paradise as one can get. He prefers marinas, seaports, docksides: the ports of departure for those other, Edenic places; the stepping-off points in our species' ceaseless drive to explore the worlds away from our homes. I suppose that was one reason for my own interest in him. It is one thing to appreciate the homogenizing impact of introduced species away from home, against a remote and exotic backdrop like the rain forests of Hawaii, where the contrast is stark. But how readily would it stand out in more familiar environs, in the semi-urban habitats that the majority of the world's people actually inhabit? I was interested too in the nature of nature. The average person living in so-called civilization is inclined to view nature as a world apart from the realm of humans—a semantic construction that, while understandable, unfortunately reduces nature to an entity that we cannot actually inhabit. "The natural world is far more dynamic, far more changeable, and far more entangled with human history than popular beliefs about 'the balance of nature' have typically acknowledged," writes the environmental historian William Cronon. "The task is to find a human history that is *within* nature, rather than without it." Jim Carlton, it seemed to me, was up to his elbows in exactly that.

16

"Here's a nice little seaweed," Carlton said. It was eight a.m., and he sat cross-legged on a floating dock in the San Leandro marina, a few miles south of Oakland on the east side of the Bay, poking through what looked like a green, waterlogged feather duster that he had just pulled up from the dock's underside. The leisure craft of the local privileged—small yachts, sloops, pleasure cruisers—sat at ease around him. The water, olive and murky, was as slate to a bright and cloudless sky. From time to time the Oakland airport indicated its proximity with the passing, directly overhead, of an enormous roaring jet.

San Leandro had been the first stop of the day. After disembarking from the van, the crew of biologists quickly fanned out to strategic points on the marina dock. Andy Cohen eyed one of the large Styrofoam buoys underneath the dock that helped keep it afloat. Just below the waterline, a bright forest of organisms had assembled itself. Cohen reached down and began ripping off handfuls of life and spreading it in the sun: dark mussels, orange sponges, small grayish oysters, delicate strands of green seaweed, saggy yellow stumps of sea anemones. Claudia Mills, from the University of Washington, had dropped a plankton net—a long cone of fine mesh, similar in appearance to a wind sock—into the water and was reeling it back up to the surface. With a handheld dredge, John Chapman of Oregon State University was hauling up mud from the estuary floor. "Claudia's looking for jellyfish," Carlton said. "John is always scuzzing for amphipods." At one point Carlton lay on his stomach and dipped a salinity meter into the murk. "Anyone want to guess the salinity?" The salt content of mid-ocean seawater is about thirty-five parts per thousand; of freshwater, zero to one part per thousand. The marina was several miles from the Pacific headwaters of the Golden Gate, so Carlton put the local salinity somewhere in the low twenties. He checked the meter: "Just about twenty on the nose."

Encompassing some sixteen hundred square miles of surface waters, the San Francisco Estuary is one of the largest freshwater and estuarine ecosystems in the United States. The estuary spans four linked embayments and the Sacramento–San Joaquin river delta. San Francisco Bay proper, which begins in the south around Palo Alto and Fremont and runs north to San Rafael and the San Quentin prison, is only part of it. Carlton's studies also include San Pablo Bay, just north of the Richmond–San Rafael bridge, and continues some way into the Sacramento–San Joaquin river delta. This broad area contains representatives of most every kind of aquatic habitat found in the warm and cool latitudes of the nation. There are fresh, brackish, and saltwater marshes, sand flats, mudflats, rocky shores, beds of eelgrass, shallow-water ecosystems on a variety of sediments. Salinity, one of the key factors governing which species of aquatic organism live where, runs from near-ocean readings at the Golden Gate Bridge down to single digits near the mouth of the Sacramento River. A few organisms, including the introduced European green crab, can tolerate a wide range of salinities. But most inhabit very narrow spectra of saltiness, which limits their potential range, although even these limits are changeable: shifting tides, rainstorms, and droughts can dramatically lower or raise the local salinity in large portions of the estuary.

On the whole, the wide variety of habitats and potential inhabitants make the region "an ideal theater for assessing the diversity and range of effects of aquatic invasions," wrote Carlton and Cohen in their summary for the Fish and Wildlife Service. Their follow-up surveys are designed to collect specimens from as many different locations as a handful of biologists can reasonably visit in a week. This year, eighteen sampling sites were on the itinerary. Many were marinas, each with a slightly different salinity. Large commercial shipyards, not small marinas, are where the greater number of foreign marine organisms are first introduced, but marinas are the better locus for an ongoing survey of the problem, Carlton said. To begin with, marinas are safer and easier to access, and limiting the survey to marinas lends a standardized element to the study. Also, not all commercial-shipyard supervisors warm to the prospect of scientific investigators swarming around the property, regardless of what those scientists might be investigating. Moreover, Carlton's main interest is in understanding how introduced organisms spread and become established once they've arrived, not simply what happens at their point of arrival.

"A lot of things don't survive past the drop zone," Carlton said. He was examining the shapeless mass of seaweed in front of him. "It may be a *Bryopsis,* introduced from somewhere. Boy these are handsome!"

Bryopsis is a broad genus of marine algae, or seaweeds. Like all weeds, the *Bryopsis* are hardy and liable to pop up anywhere: most visibly on natural or man-made floats, or at the low-tide mark of wooden pilings. They were common hangers-on in the days of ship fouling and have traveled widely as a result. The strands of seaweeds are themselves a movable feast, host to any number of microscopic organisms that feed, live, and incidentally ride on their edible surface. What those subvisible organisms might be, Carlton would later determine in the lab; for now, he put the specimen of *Bryopsis* in a jar of seawater. Meanwhile, the other biologists were coming and going on the dock, presenting Carlton with their own discoveries. Dislocated life-forms accumulated in a heap around him: a cluster of Japanese mussels, each hardly larger than a fingernail; the calcified burrows of a colony of tube worms, which rather resembled a small, crusty pipe organ; globs of sponges. Most of the organisms arrived not in single units, but in tight-knit, intermixed clumps: barnacles on mussels, seaweeds on barnacles, everything clinging tightly to everything else, with tiny isopods, like pill bugs, trundling around in it all. Carlton, handling one such clump, inspected what looked like a row of small, glutinous ball bearings.

"Most of these yellow things are anemones," he said. Removed from the water, the anemones had withdrawn their tentacles and turned inward to conserve moisture. "*Haliplinella luciae.* They're Japanese. They first showed up in Rhode Island in the 1880s and '90s." From across the dock, the other biologists were shouting out the names of the organisms they were pulling up: *Mytilus galloprovincialis, Schizoporella unicornis, Molgula manhattensis.* My skills suited me to the task of writing down all the scientific names as they were called out. At one point Carlton pried apart an encrusted, living mass of exotic matter and uncovered, at the bottom of everything, a small whitish oyster. "What's native here? This tiny oyster, that's about it."

The San Francisco Estuary has received so many introduced marine organisms largely because it has for so long attracted human mariners. Propagule pressure. The estuary is a vast, sheltered, and penetrating waterway; the sea-lanes stretch from the coastal headlands to the inland port cities of the Central Valley: Pittsburg, the outskirts of Stockton. The port of Oakland—a sprawling workscape of cargo ships and gargantuan cranes, the alleged inspiration for the sinister, oversize combat vehicles in *Star Wars*—is by some measures the largest in the United States. The nation's last whaling station, a blood-rust carcass of a building, sat on a lonely stretch of shore on San Pablo Bay until it was finally dismantled in 1998. The first ship known to have entered San Francisco Bay, the *San Carlos*, arrived in 1775; ships soon flocked there from both the Atlantic and the Pacific oceans. By the early 1800s the estuary facilitated a steady export of lumber, hides, and furs; with the 1848 gold rush, the region blossomed overnight into an international center of shipping and trade. The native creatures in the estuary had always been subject to the vicissitudes of nature: currents, waves, winds, and storms—seasonal, annual, or sudden—that altered the local salinity or eroded the sediment of their habitat and sent it elsewhere. And their numbers may have been thin to begin with, Carlton believes; the estuary is only seven thousand years old, relatively young in ecological terms. Now an additional array of disturbances came into play. Rivers were diked; channels were dredged; marshes and mudflats were filled. Dams rose up, rivers changed course, shorelines were redirected, and sediments and toxins poured in from new sources: farms, mines, refineries, factories, ships, cities. New habitats were etched—seawalls, pilings, riprap—and old slates wiped clean.

Most occupants of an ecosystem as fluid and abruptly changeable as an estuary are accustomed to disruption: a mudflat might be washed lifeless, but enough native mud crabs and native mud snails are floating around in larval form that sooner or later, somewhere, they settle down and the population of native mud crabs and mud snails rebuilds itself. After 1848, however, the disruptions in San Francisco Bay became ever more frequent and abrupt, and each new tabula rasa made way for a novel array of exotic contenders intent on—and adept at—eating, settling in, and spreading out. With the completion of the Transcontinental Railroad, in 1869, East Coast oysters were introduced to the Bay along with a host of oyster hitchhikers. "A single oyster shell may have upon it representatives of ten or more invertebrate phyla, comprising dozens of species," Carlton has written, "and these numbers can be greatly increased when oysters are packed together for shipment with associated clumps of mud and algae." In the subsequent four decades, Atlantic oyster seed was sent west and planted by the millions. Pollution and the changing hydrology of the Bay shut down the industry after the turn of the century; a minor industry based on the Japanese oyster thrived briefly in the 1930s. By then, Carlton says, the incidental companions of oysters had already made themselves apparent. By the 1890s several clams and snails common on the East Coast were recorded in the Bay: an oyster drill, a gem clam, a marsh mussel, two species of slipper limpets, a mud snail. In 1946 the Japanese mussel *Musculista senhousia* and the Japanese clam *Venerupis philippinarum* were collected. Of all the introduced marine organisms Carlton has found in the Bay so far, 15 percent—some thirty species—came in with Atlantic oysters, he figures; the Japanese oyster is associated with another 4 percent. In describing the introduction of marine species, marine biologists sometimes resort to a hypodermic analogy. For a century and a half, writes Carlton, the human traffic has subjected the Bay Area "to both multiple and massive inoculations of exotic species." Today San Francisco Bay is home to more species of exotic invertebrates than anywhere else on the West Coast.

By three o'clock Carlton and company had made their way a few miles farther up the east side of the Bay, to a marina operated exclusively by and for the U.S. Coast Guard. Here was another floating dock, one much smaller than the last, upon which the biologists again threw themselves to extract nature and its information: temperature, salinity, seaweeds and sponges and mussels detached from pilings with an audible rip. "Here's a flatworm," Carlton said, noting one squirming deep in a knot of dislodged sea life. Flatworms, he conceded, are not glamorous subjects; consequently, there are fewer flatworm biologists than some might wish. "There's a whole career in there; the world would beat a path to your door." At one point he and Cohen located a large, flat rectangle of Styrofoam floating loose in the water alongside the dock. With some effort they turned it over. The underside was a thick slime jungle, a gloppy, multicolored mat of sponges, sea squirts, and bryozoans. Carlton and Cohen pulled off a few specimens for their sample jars, then tried to turn the float right side up. Its large size, however, and the fact that 90 percent of its mass was now on top instead of underneath, made it impossible to right. After several minutes of futile effort, they gave up. "Uh, let's just walk away whistling," Carlton said.

Nearby was a narrow, sheltered beach. Carlton and Cohen walked down to it and began to stroll its length. They flipped rocks over with their feet, so I did the same. A snail caught my eye, and I picked it up. It was small and whitish and looked pretty much like every beach snail I've ever seen—with good reason, Carlton said. "That's *Littorina saxatilis*, the Atlantic periwinkle. It's the common snail of the East Coast." The snail grazes on algae and has the potential to drastically—if subtly, to the average human eye—alter the intertidal environment. The overall effect might be not unlike the introduction of a herd of tiny, slow-moving aquatic goats. In 1993 Carlton discovered the first population of *Littorina* on the West Coast, at the marina in Emeryville, next to a public boat ramp. He strongly suspects that it arrived hidden in the seaweed in which live bait-worms are packed and shipped from Maine to San Francisco; fishermen habitually toss the seaweed from their boats into the water. More than once, Carlton and Cohen have casually inspected shipments of bait-worms and found live snails in the seaweed; as many as a million bait-worms are shipped to the area each year. Carlton and Cohen had collected more than a hundred specimens of *Littorina* in Emeryville. "We've been watching the population, wondering if we should eradicate it," Carlton said. It so happened that the specimen I'd picked up was the first they had seen anywhere other than Emeryville. Already Carlton had mentally sketched out a scientific paper in which he describes a second, newly established population of *Littorina* in the Bay. "Find two more of these, and you'll be a coauthor," he said.

An invasion biologist is frequently in the awkward philosophical position of admiring an organism that, by all accustomed measures, should at best be maligned. An introduced species may be a nuisance, even a menace, but it is also, in a sense, a winner, and even a biologist can't help but be impressed, in a scholarly sense, by a winner's ability to survive. "And not only survive but become phenomenally abundant," Cohen said. "I think everyone who works in this field is impressed by that. That's part of what draws people to it. We root for these things all the time, though we don't admit it."

"Yeah," Carlton said. "I got excited when I saw that second population of *Littorina saxatilis* in the East Bay. I'm sorry to see it established here; I suppose I should have had a little wake or gone out with some flame-throwers. Yet as an ecologist, it interests me that it is getting established and how it will spread—and how difficult that is to measure. I mean, it would be an incredible labor of love to figure out where *Littorina saxatilis* is right now. Its distribution is very patchy; they're hidden among rocks, in the dark. It's fortuitous how we found them. That means there are probably a fair number of little colonies of them kicking around."

"And it's unlikely that anyone else will notice it, or notice that anything is different," Cohen said.

"It's probably in a fairly early colonization stage," Carlton said. "We have no idea how abundant it will become in the Bay."

They made their way back to their colleagues, who were still dredging off the docks; then everyone tallied their gains and piled back into the van. If they limited their time to two hours per site, including driving time, the group would hit its goal of visiting nearly two dozen sites in the coming days: docks, wharves, boat slips, and pilings, each site a slightly different habitat from the next, each with its own suite of exotic marine slimes, encrustations, seaweeds, and worms to rip off, pull apart, and note down.

"Shipworms! I got shipworms!"

"I've got mysids!"

"Who wants this flatworm?"

"Here's a baby *Hemigrapsus*."

"That's not *Hemigrapsus*, it's *Pachygrapsus*."

"What's that gunk over there on the dock?"

"It's dried gunk."

"This sieve is rusty. It's great except for the big hole."

"I didn't get a grant to pay for it, like you guys do. I bought it on sale."

"Where, at KMart?"

At the larger and ritzier marinas, Carlton or Cohen would spend a few initial moments chatting up the dock supervisor while the other biologists got to work. Others began with a surreptitious climb around a chain-link fence and ended with a warning call from Carlton: "Work with alacrity, folks. We may be kicked off soon."

Channel Street is a two-lane thoroughfare along the coast at the southern edge of San Francisco. A grassy ridge parallels the western side of the road; beyond it is sand, then the Pacific. In the late nineteenth century, Channel Street was an open waterway, one of several that carried raw sewage from the city out to sea. This channel in particular, a health officer wrote at the time, "smells to Heaven with a loudness and persistence that the strongest nostrils may not withstand and the disinfectants of a metropolis could not remove." Today the Bureau of Water Pollution Control has its headquarters on Channel Street. From the road, the only visible aspect of the bureau is an enormous corrugated-steel garage door built into the hillside, and a security telephone. Carlton used the phone to dial a guard somewhere inside; presently, the door was raised and a warning siren blared. He steered the van through a concrete tunnel into a central concrete parking lot surrounded by high concrete walls. The place was vast and ghostly, even in daylight, and had the air of a heavily armed fortress.

A certain fraction of the Bay's aquatic inhabitants were scientifically identifiable with the naked eye as Carlton and his colleagues pulled them from pilings and floats and piled them on the dock. But most were too small, or their defining characteristics too intricate, for even an experienced marine biologist to quickly tell what was what or to say which species were native to the Bay and which were newcomers. So for the next few nights and through the weekend, the bureau laboratory would be occupied by Carlton's survey team and its microscopic preoccupations. Andy Cohen had set up the arrangement. The lab was spacious and sparkling clean, with wide workbenches, comfortable stools, and cleaning areas with steel sinks and spray-nozzle hoses. When I caught up with the expedition one afternoon, the crew members were spread throughout the room, each peering into a microscope at some teaspoon fraction of any one of the dozens of sample jars of specimens collected earlier in the week.

With some excitement, Carlton motioned to his microscope, where he had an isopod in focus. I looked through at what resembled a fourteen-legged armadillo wandering through a moonscape of barnacles and collapsed anemones. Isopods are the marine equivalent of potato bugs; this one belonged to the genus *Sphaeroma,* although Carlton had not yet identified the precise species. Whatever it was exactly, Carlton previously had seen it only on the east side of the Bay; this year, for the first time, he had found a population on the west side. "I should be able to get a species name on it when I get back home," he said. "Oftentimes you only get one or two specimens, so you're not sure whether to make a big deal of it." Usually *Sphaeroma* is found in wood pilings, in the mud, or in the vacant tunnels of shipworms. Carlton has discovered that on the West Coast, the isopod has a tendency to burrow into Styrofoam floats and buoys, for which reason he has begun referring to the genus as *Styroma.*

Across the room, Cohen had a bryozoan in focus. "A bryozoan" in fact is a tight-knit colony of individual bryozoans, or zooids, each one little more than a tiny lump of digestive tract in a calcareous box. From each box, a small fan of tentacles waves about and draws food in. Technically the bryozoan is an animal. Bryozoans grow in clusters of thousands of identical units, an interconnected network of food processors. Bryozoan colonies assume various forms, depending on the species. Some are flat and reticulated and can blanket broad portions of rocks, floats, and wooden pilings; to the naked eye they somewhat resemble lichen. Others assume a more upright, branched form and resemble (under a microscope) dense forests of greedy, grasping trees. (One textbook informed me that "many people have erect bryozoans in their homes without knowing it.") The bryozoan colony Cohen was looking at was of the branching variety, and a tiny mat of it had settled around the even tinier branches of a sponge. Carlton walked over and peered in. "This is a mess, isn't it?" he said, impressed. "It's like living in a thornbush. You wouldn't think space is that valuable, would you? If they had voices, they'd be going, 'Ooo! Aah! Oow!' "

The study of marine invertebrates, even more than the study of terrestrial invertebrates, demands a unique combination of patience, devotion, and strength of eye. Yes, the soil of the earth is filled with countless species of subvisible microbes and insects. But soil is only the thinnest, outermost layer of Earth's skin. In contrast, the sea is miles deep and thousands of miles wide; it covers 71 percent of the globe's surface, its total volume is eleven times greater than all the land above sea level combined, and virtually every cubic inch of it is teeming with invertebrate life—much of it microscopic; most of it unseen or unidentified; most of the remaining, fractional portion that has been retrieved and observed comprised of fragile, translucent, and dazzlingly intricate organisms seemingly indistinguishable from every other species of fragile, translucent, and dazzlingly intricate organisms. Carlton said, "There are people who look at invertebrates by habitat, others by a strictly taxonomic orientation, others with a broader ecological context, like I do with introduced species. But it seems to me, when they fall in love with an invertebrate group, they're fascinated by the morphology. You learn about the taxonomy; a great number of species are not yet described. It all becomes great fun. But unless you're a museum systematist, it's something of a hobby."

In the eighteenth century, the Swedish botanist Carolus Linnaeus laid out a scheme that classified the world's life-forms into groups, or taxa, according to the physical shapes of their bodies. A century later Darwin's notion of evolution was added to the mix, such that today the Linnean system, while still concentrated on morphology, classifies organisms based on their ancestral relationship to one another. Toward the end of the nineteenth century, however, the German biologist Ernst Haeckel proposed an alternate but complementary classification schema—still in widespread use today—for the marine organisms, based not on how an organism looks, but on where it lives. Haeckel divided the sea's inhabitants into two groups: the benthos (Greek for "deep"), which live on or in the seafloor; and the nekton (from *nektos*, "swimming"), which encompassed everything else above the benthos. Isopods, anemones, flatworms, bryozoans, horseshoe crabs, clams, limpets, corals: the benthos are sedentary creatures, often immobile, in any case limited mostly to a territory of two dimensions. Fish, sharks, squid, whales, and all things swimming are nekton. By subdivision, later scientists created a third category: the "wandering" (*planktos*) plankton. The plankton are those organisms—mostly but not exclusively quite small—that do not swim freely as the nekton do: krill, seaweed seeds, the drifting larvae of sea snails. Although some plankton possess tiny appendages with which they flail against the water, overall their travels are determined by the drift of tides and currents. The distinction boils down to locomotive ability: nekton can move willfully against a current; plankton are at the whim of the flow. It is the difference between a mariner and a castaway, an ocean liner and a lifeboat, a voyaging canoe and a balsa wood raft. Forget the countless species of terrestrial insects: by far the most abundant form of biological matter on Earth is plankton—particulates of life mostly, going nowhere in particular and in no particular hurry.

These three categories—nekton, plankton, and benthos—span the boundaries of Linnean classification. For example, the plankton include members of both the plant kingdom (phytoplankton) and the animal kingdom (zooplankton). Members of the zooplankton include everything from jellyfish (phylum Cnidaria) and their cousins the comb jellies (Ctenophora), to shrimps (Arthropoda), some snails (Gastropoda), all the way down to single-celled protists of the genus *Globigerina.* Picture the *Globigerina* as amoebas encased in fragile shells, which they manufacture from cemented grains of sand or from calcite they absorb from seawater. Upon the death of the animal, the shell drifts to the ocean floor to join a thick, billowy layer of micro-shell muck known as carbonate ooze. The ooze accumulates at an approximate rate of two inches every thousand years; scientists estimate that one-third of the ocean floor is covered with a layer of carbonate ooze anywhere from half a mile to three miles thick. The chalk cliffs of Dover are comprised largely of ancient, uplifted beds of carbonate ooze—mountains built on microscopic backs. Among phytoplankton, the tiny dinoflagellates are perhaps most common. Each consists of a single cell, with a handful of chloroplasts for photosynthesis and two tiny flagella to whirl around with. They are like plants with tails; some are elaborately armored with horns or spines. If conditions are right, dinoflagellates divide and multiply into massive blooms that turn the water red or green or brown. Dinoflagellates and diatoms—single-celled plants, many with no means of propulsion—constitute the bulk of plant life in the sea. Science books often refer to them as "crops" and "pasturage" on which the rest of sea life grazes. In the marine food chain, they are most of the food.

Of the three categories, that of plankton is the most plastic. Some plankton are simply nekton in transition: the larvae of many fish may spend days or weeks drifting as plankton. A great many benthic invertebrates—sponges, anemones, corals, mussels, crabs, barnacles, sea stars—cast planktonic larvae into the currents. Many phytoplankton are merely the seeds of seaweeds and stationary sea plants. Some species (known collectively as holoplankton) spend the whole of their lives in planktonic form: krill, for instance, and the transparent arrowworm. Others (the meroplankton) devote only a portion of their allotted time, usually the larval part, adrift in the water column. The jellyfish is one. A few species of jellyfish are incapable of swimming; the adults, which have no tentacles, creep around on mats of algae and underwater stones, eating whatever they can find. The two hundred or so other species are more familiar. The adult medusae waft through the seas, all gelatin and nettle. When the time comes, they produce sperm and eggs, and the fertilized conjunction—the planula, shaped like a microscopic sausage—settles and forms a small polyp, or scyphistoma. Under a microscope, the polyp looks rather like a blossom, with waving petals that herd food particles toward a central opening, the animal's mouth. Over time, the body of the polyp changes into a column of individual disks; one book I read compared the formation to a stack of hotel ashtrays. One by one, the disks are cast off. These are young jellyfish, free to live the meandering life.

Claudia Mills was looking at something like this through her microscope. Mills, a member of the annual Bay expedition since its inception in 1993, is an expert on jellyfish and their cousin comb jellies. The present specimen, a young comb jelly, had come from a marina toward the southern end of the Bay, where the team had gone at six o'clock that morning. At the moment, the animal was in its rooted form, growing at the top of what looked like a stem, and went by the species name *Ectopleura crocea.* It was an astonishing little creature, a translucent red daylily, winking. It too was an invader, noted in the Bay since the 1850s, but like so many alien species, its potential impact was unclear. "It looks as much like a flower as an animal can look," Mills said. "It's something we expected to find at several sites, but we only found it at two."

The biologists that Carlton and Cohen assembled in 1993 were invited based in part on their invertebrate specialties. Their combined expertise would make the study of impossibly small and seemingly indistinguishable organisms at least marginally easier. John Chapman, from Oregon State University, volunteered with a deep knowledge of small crustaceans: isopods, mysids, cumaceans. Leslie Harris, from the Los Angeles County Museum of Natural History, knows the polychaetes, a diverse class of bristled, segmented sea worms that comprises more than five thousand species. She drew my attention to a specimen she had identified in one of the Bay samples. Its skin was a translucent red, tinted most likely by the algae the worm feeds on, Harris said. Through its skin I could clearly make out a single blood vessel that ran the length of the worm; a welt of blood pulsed from one end to the other, steadily as a metronome. In recent months Harris has taken an interest in a South African species of polychaete that lately has devastated California's abalone industry. The worm bores into the shells of living abalone; it does not harm the animal nor damage the meat. However, the shells of afflicted abalone are left misshapen—"like baseball caps," Harris said. Alas, baseball caps do not sell well in the seafood section; sales of abalone have plummeted.

Suddenly Carlton called out from across the room, "I've got *Corophium* here!"

It was *Corophium brevis,* a tiny, shrimplike amphipod that serves as a food source for many fish. It was a Bay native, which seemed to explain Carlton's excitement. Still, his announcement failed to elicit a response from the other biologists, absorbed as they were in their own microworlds.

"Nobody's listening," Carlton said to the room at large. "I could have thousand-dollar bills here, but nobody's listening."

Carlton had two microscopes set up in his work area. He shuttled back and forth between them, slipping glass petri dishes under one lens or the other and taking note of the residents: a spionid worm; a diminutive hermit crab; some small, shapeless snail larvae that Carlton described, fittingly, as "blobby goobery." He inspected the movements of a flatworm, a spotted, amoebalike mass as large as a nickel. Two small eyes rose on stalks at what was, apparently, the front end. Under the spotlight of the microscope, the flatworm glided across the petri dish with an elegant fluidity, like some hybrid of a flounder and a leopard-skin rug. Carlton next turned his attention to a pycnogonid. "Ever seen a sea spider? They're called that not because they're spiders, but because of their general morphology." The pycnogonids comprise one of several classes under the arthropod phylum, which also includes the crustaceans, insects, and arachnids, or true spiders. The pycnogonid Carlton had in view did indeed look spidery, despite the distant relationship. On ten long legs it proceeded gingerly across a potholed microscape, like Charlotte on rickety knees. "I don't know which sea spider it is, frankly," Carlton said. "But it's a classic fouling organism. It's deeply embedded in communities that are otherwise introduced."

In a landscape as confined and confounding as that of marine invertebrate biology, populated as it is with mobs of transparent, multi-legged blebs that might be the planktonic stage of just about anything, it helps to have directions. Enter the identification key. Over the years, taxonomists—specialists with an in-depth knowledge of a particular group of organisms—have assembled a variety of manuals, or keys, to help the less-specialized biologist identify whatever odd or unfamiliar plant or animal presents itself under the microscope. Dozens of such books were on the shelves in the bureau laboratory, and every member of the Bay survey team had one or more propped open on his or her workbench. *Intertidal Invertebrates of California. Hydromedusae of British Columbia and Puget Sound. Pacific Coast Pelagic Invertebrates: A Guide to the Common Gelatinous Animals.* One guide opened on a bluntly honest note: "The identification of marine plankton is a tedious yet fascinating undertaking."

A key such as this functions as a guidebook: you compare the unknown organism in front of you to a series of increasingly precise descriptions of known organisms in the book, following along until you achieve a positive identification. Is your organism "elongate, cylindrical or chain-like"? Does it have "spines about twice the body length, not widely spreading"? Then it must be the echinopluteus larva of a sea urchin. Does it instead possess "spines more than twice the body length"? Well, then, that's an ophiopluteus larva of a brittle star. If it has "ciliated lateral bands, leaf-like in appearance," then you are looking at the auricularia larva of a sea cucumber. If it has "tentacles or developing feet," that's the brachiolaria stage of a starfish.

For the purposes of deciphering the contents of San Francisco Bay—what's native, what's introduced—by far the best general manual is a blue hardback book entitled *Light's Manual: Intertidal Organisms of the Central California Coast, Third Edition,* which represents the edited insights of dozens of individual taxonomists. Even Carlton was consulting a copy, although one might think he would already know everything in it: he edited it. The first edition, assembled by an invertebrate biologist named Sol Felty Light, was published in 1941 and titled simply *A Guidebook to the Intertidal Invertebrates of Central California.* The second edition appeared in 1954 and had five editors, including Ralph Smith, a zoologist at the University of California at Berkeley. In 1973 the aging Smith encountered a promising young marine biologist named Jim Carlton and invited him to coedit the third edition. The fourth edition, incorporating the insights of more than a hundred taxonomists on the Pacific Coast, is in the works. Carlton is the sole editor; he plans to rename it the *Light and Smith Manual,* in honor of his former colleague, who died in 1993. For some time now, Carlton has been meaning to assemble an oral history of the aging taxonomists, to record their accumulated knowledge, but so far he has not gotten around to it. "Every time one of these geezers dies, I feel terrible."

Cohen's voice came from the other end of the room. "Have you identified that *Limnoria,* Jim?"

"I'm working on it."

One of Carlton's microscopes had in focus a chunk of decaying wooden piling collected from a floating dock near downtown San Francisco. The wood had been gnawed heavily by an isopod of the genus *Limnoria,* also in view. It was white and fat and grublike; Carlton had yet to determine its precise species. His attention was momentarily drawn to the wood itself and to several tubular boreholes that ran its length, one of which held a slender white shipworm. Rough handling had torn open its body like a peapod: inside was a series of what looked like tiny white clams. "We've broken open the brood chamber," Carlton said. "Here's what shipworms look like in the planktonic stage. Very few people in the world have seen this."

He focused again on the isopod *Limnoria*. "They're known as the termites of the sea. Probably he's not doing so well, as his head is partly crushed. These are an introduced species, another one of our Flying Dutchmen. I don't know the origins, but it wasn't here in early records. Look at all these galleries." He directed my attention to a network of small grooves in the wood gnawed out by the isopods. "They create the classical hourglass shape on dock supports. They bore from the outside in, whereas shipworms bore from the inside out. Look at that little *Limnoria*, looking up at us with his black beady eyes."

To Cohen he said, "It's either *Limnoria tripunctata* or *Limnoria quadripunctata*. It could be *lignorum*."

"I thought we never saw *lignorum* in the Bay."

"Oh, maybe around the Golden Gate."

Carlton looked back into the scope, then directed me to look again closely. From the hind end of the isopod extended what looked like the open ends of several tiny green tubes or pipes. Amid the tiny tubes grew tinier stalks, each with a globule at the end, like a pinhead. From time to time, from the end of the green tubes at the end of the *Limnoria*, minuscule horseshoe-shaped probes popped briefly into view, then disappeared again. "These *Limnoria* have a little commensal organism on their backs, called Mirofolliculina," Carlton said. "It's a protozoan that lives on these things—just on *Limnoria*, as far as I know. Are those cute or what?" The pinhead stalks, he said, were yet a third organism altogether, another form of protozoan. "And midships, on the left side of *Limnoria*, there's a little discontinuity—that's a commensal copepod." Sure enough, I could make out a translucent lump, vaguely reminiscent of a crustacean, firmly attached to the isopod; it wasn't parasitic, merely adapted, barnacle-like, for a permanent ride. "There's a paper in German about commensal copepods on *Limnoria*," Carlton said. "This one hasn't been recorded before in the Bay."

Cohen came over and took a look. "That looks uncomfortable, all those things growing out of your butt."

"It's a veritable fuzzball of commensals," Carlton said. "The rats have fleas, which have smaller fleas."

17

The Bay survey team includes on its itinerary of sampling sites Lake Merritt, a small, brackish lagoon surrounded by a public park smack in the center of downtown Oakland. In 1870 the lake was established as a wildlife refuge, the nation's first. A murky pond amid dull greenery, with an empty parking lot and several nearby office towers for company, Lake Merritt today is more a lunchtime refuge for lawyers than a haven for threatened fauna. Nonetheless, it holds special interest to marine invertebrate ecologists, one in particular. "There are a lot of introduced species here," Carlton said. "Originally it was a slough attached to the Oakland estuary. Since then it's been a highly urbanized system." The team of biologists, having tumbled out of the van, had taken up stations along a floating dock that extended out into the lake. Carlton sat out at the end, sifting through a soggy heap of living matter he had pulled from one of the supporting floats. He yawned and stretched and looked out over the water, which was green with algae. "Boy, is this water productive or what? The phytoplankton are just *blooming*."

For all intents and purposes, the modern study of marine biological invasion began on a September afternoon in 1962 during a Carlton family picnic to Lake Merritt's meager shores. Jim Carlton was fourteen. Hoping to expand his seashell collection, which he had been accumulating since age nine, he wandered down to the lakefront. He jumped from a small bluff down to the beach and landed on something hard: a clump of crusty organic tubes, what looked like a petrified hornet's nest, but obviously marine in origin. Later Carlton would come to know it as the characteristic housing of tube worms, which live underwater in dense clusters of slender tubes that they manufacture from calcium carbonate. At the time, however, the young collector did not know what he'd found, so he took it home and showed it to his mother, who promptly relegated it and him to the basement. The following week, he wandered back to the park's nature center and sought enlightenment in the exhibits. His attention was drawn to a specimen identical to his and labeled *MERCIERELLA ENIGMATA*. "It was misspelled: it's actually *Mercierella enigmatica*. Today it's called *Ficopomatus enigmaticus*." The two words that seduced Carlton that day, however, were written below the species name and described the native origin of the tube worm: SOUTH SEAS.

How had it arrived there? What else in the lake was foreign? Beginning right away, until 1974, Carlton conducted a twelve-year survey of the biological contents of Lake Merritt. He established ten sampling stations along the margin of the lake and visited them weekly with specimen containers. He built a laboratory in his parents' basement, complete with microscopes and ten burbling ten-gallon marine aquariums. With the generous help of various old-time systematists at the California Academy of Sciences, he taught himself invertebrate biology and taxonomy. "I first wandered over there in 1962, when there were folks who'd begun their careers in the 1910s and '20s. Almost all of them are gone now. From the age of fifteen on, I began visiting on a regular basis, taking the bus over, dragging my species with me. The academy didn't have sponge people, hydroid people, polychaete people. But they taught me how to send my species to experts around the world: how to box them, where to mail them. I was shipping stuff to Canada, to Germany. Everyone helped me. They'd write back, 'Dear Dr. Carlton . . .'" Many of the responses also included the phrase "No previous record on the Pacific Coast." Again and again the adolescent Carlton was discovering species in Lake Merritt that by all biogeographical logic should not have been there. His success rate was a function of sleepless diligence and his incidental choice of a nonclassical environment. "Nobody had been collecting in the brackish, estuarine areas around the Bay. Who'd want to muck around in the backwaters? Being young and naïve, I tackled everything: nematodes, polychaetes. I wanted to know it all, no matter how big or small." He wrote his first scientific paper, on the fauna of Lake Merritt, for the *News of the Western Association of Shell Clubs*, at the age of fourteen. Many of the species Carlton described then are still thriving in the lake, including a small Asian shrimp that Carlton—now peering down into the water at an active specimen—had suddenly taken an interest in. "It was introduced by a ship from Korea in the 1950s, during the

Korean War. I used to catch it as a kid, when it must have just arrived. It pops up now in places like Coos Bay, in Oregon."

The exotic mysteries of Lake Merritt were merely a hobby, a diversion largely unknown to young Carlton's friends. In junior high and high school, his overt passion was language: English, journalism, literature, the humanities. He was editor of the high-school newspaper. He interviewed Jack London's daughter Joan, then in her sixties, for the school literary magazine. He considered studying linguistics at Berkeley. As well as a basement laboratory, Carlton kept an office upstairs in his parents' house where he was slowly assembling his first, yet-unfinished monograph—an exhaustive collection of English-language clichés: *white as a . . . green as a . . . red as a . . .* He scoured books that his mother, with a master's degree in linguistics and literature, had accumulated. "I was interested in the origin and evolution of slang words, in the differences in regional dialects. I liked the idea of a fingerprint, that I could tell where you'd grown up, whether you'd spent a summer in Missouri." He was the world's only linguodeltiologist—a word and professional specialty he had invented to describe himself. Deltiology (a real word) is the collection and study of postcards. Scouring the junk shops of Oakland's lower east side, the young Carlton turned up box after box of postcards dating from the 1870s onward, vacation postcards mostly: Coney Island, Florida, Palm Beach. He sat on the floor and studied them, looking for slang words, noting each card's date, its origin, the destination address. "Here was a language; here was oral history. Words were written down that weren't likely to appear in print, at least not in 1900 or 1910. I'd find a slang word—*car* for 'auto,' or a racial epithet. I'd look at the date; I'd look in my mother's *Dictionary of American Slang.* It would say that the first usage was 1927—but here it was being used in 1917. I'd found an earlier history. I had a window on linguistic history." During his high school years, Carlton took no chemistry, no physics, no biology. Compared with the professionals he'd been quietly mingling with at the Cal Academy, the high school science teachers had nothing to offer him. "My interests came as quite a shock to the science nerds."

Eventually, science won his full attention. Carlton attended the state university in Berkeley and, in 1971, received his bachelor's degree in paleontology—"which was a guarantee for absolutely nothing." Throughout college he had worked as a volunteer at the Cal Academy. On the Monday after graduation he was offered a full-time job as a research assistant in invertebrate zoology, for thirty-six hundred dollars a year. "I was in heaven. I could save up some money, buy a car. I had no particular plans after that." That year, two Standard Oil tankers collided under the Golden Gate Bridge, generating a spill that seeped well into the estuary. Carlton was asked to sift through the biological wreckage and help identify some of the smaller invertebrate specimens. On the job he met John Chapman, then a graduate student in biology at San Francisco State University, who also had been called in on invertebrate duty.

"Amphipods are about all I work on now," Chapman said. He was working at the end of the dock now, alongside Carlton. Amphipods are crustaceans. They resemble, more or less, minuscule shrimps. With tweezers, Chapman picked individual wriggling specimens from a mass of vegetation from the bottom of the lake. "They're a great animal to do biogeography with. They brood their young. They don't have planktonic larvae, they don't drift far, so it's easy to tell what's from where. I've seen two species arrive in the Bay since Jim and I began working on amphipods. One of the advantages of working so long in a place like this is when you see something new, you know it wasn't there before. Many people didn't want to believe these things are introduced, but it'd be almost impossible to explain any other way."

In 1975 Carlton coedited the third edition of *Light's Manual* with Ralph Smith. With the manual on his résumé, Carlton had no trouble winning acceptance to the graduate ecology program at the University of California at Davis. What began as a teenage fascination with Lake Merritt expanded into a nine-hundred-page Ph.D. dissertation identifying more than three hundred introduced marine invertebrates on the Pacific Coast, including nearly a hundred exotic species in San Francisco Bay. Carlton put in three years as a postdoc at the Woods Hole Oceanographic Institution, in Woods Hole, Massachusetts, on the elbow of Cape Cod. He revisited an abiding curiosity: to what extent was ballast water a vector of exotic species? Carlton won a small grant to pursue the question. He began visiting the local shipyards, sampling their tanks, seeing what he could dredge up. He sampled the ballast tanks of the Woods Hole research vessels, and he rode one of them, the R/V *Knorr*, from Scotland to Newfoundland, taking periodic measurements of the temperature, salinity, and creature contents of the ship's ballast water. One day, for my benefit, he extracted from the depths of his overstuffed library a posterboard presentation of the *Knorr* experiments that he had made some years earlier, complete with a photo of him: young, curly hair not yet gray, peering up from inside a ballast hold. He rode an oil tanker from Portsmouth, New Hampshire, to Corpus Christi, Texas, sampling its ballast tanks along the way. "It was a fantastic trip," Carlton says. "I set up the microscope in the mess hall, and these old gruffs would come over and look in at the copepods."

In subsequent years, dozens of similar transoceanic sampling expeditions would be conducted by biologists—in many cases, by Carlton's postdocs or graduate students, or by their eventual postdocs and grad students. But Carlton's was the first of its kind, the inspiration for a ripening, well-financed field of science known today as ballast-water ecology. At the time, however, the subject was not yet a magnet for research dollars. Carlton moved to the Oregon Institute of Marine Biology, where he supported his ballast-water studies—including his now infamous inspection of ships arriving from Japan—with more traditional biological studies. Ostensibly, his primary research involved a study of Japanese eelgrass, a brackish weed that was gradually invading the tidewaters of Coos Bay. One morning, in the eternal hunt for research dollars, Carlton visited the director of Oregon Sea Grant, a state branch of a federal agency that allocates money to marine studies. Until Carlton entered his request, no researcher from that lab had ever won funding from Sea Grant.

"I was gonna put together a whole thing on the invasion of Japanese eelgrass in the Pacific Northwest," Carlton recalled late one evening in the laboratory in San Francisco. The Oregon Sea Grant director listened patiently for an hour, then remarked that eelgrass researchers had had difficulty attracting funding lately. Nonetheless, he added, Carlton was welcome to give it a shot. "As I was getting out of my chair and he was getting slowly out of his chair, I said, in departure mode, 'Well, yep, thank you very much, and by the way, I got on a ship recently and sampled the ballast water and found all this stuff coming in from Japan.' And he paused—he was putting on his coat; he had a lunch appointment with his wife—and he looked at me and said, 'Well, now what's that all about?' He thought ballast water was the neatest thing since the telephone. So then we walked to the elevator and he got in, and I said, 'Ahem. Which should I write the grant on, Japanese eelgrass or ballast water?' And as the door was closing, he said, 'I can't tell you that. But we've never heard of ballast water before . . .'"

It took an overactive, underfinanced ecologist several years to convince the scientific community that shipboard biological invasions, particularly those mediated by ballast water, were a subject worthy of deeper investigation. It took a brainless, striped, pistachio-size mollusk—*Dreissena polymorpha*, the zebra mussel— to bring the subject to national attention.

The zebra mussel is a freshwater bivalve. In size and coloring it resembles a small nut; a pattern of brown and white stripes on its shell accounts for its name. The animal was first identified in 1769 in Russia's Ural River. Gradually over the next two centuries it spread to western Europe: through newly opened canals and inland waterways, attached to timber imported from Russian rivers, and—so one biologist speculates—with the retreat of Napoleon's troops from Moscow. It reached Scandinavia in the 1940s, Swiss lakes in the 1960s. In 1988 a specimen was found in Lake St. Claire; a few months later another was found in Lake Erie. By 1993 the zebra mussel had colonized all the Great Lakes and entered eight river systems including the Hudson, Ohio, Mississippi, and Arkansas. Its range now extends south to New Orleans and west to Duluth, Minnesota. On a scale unlike any aquatic invader to come before, the biology and travel potential of *Dreissena polymorpha* attracted the attention of industry representatives, average citizens, local politicians, and, ultimately, Congress, the members of which hurriedly called Carlton for insight.

"Zebra mussels were the catalyst," Carlton recalled. "It was the motivational species." He stood now at the outer end of San Francisco's Pier 39 in the afternoon sun. With his back turned to the gleaming Bay, he surveyed the bustle of tourism that separated him from shore, a wharf's worth of curio shops, hot-dog vendors, ice-cream stands, and a wax museum. He was talking to Cohen. The biologists were sprawled on the dock of a marina at the extreme end of the pier, behind a padlocked gate; I had lost track of the number of sampling sites we had visited by then. The scientists went at this one with their penknives and specimen jars. A cluster of spectators watched from a distance. Carlton asked Cohen, "Any zebra mussels?"

He was half joking. Although the mollusk might well make it to the Bay Area eventually, it would never survive this close to open ocean, in highly saline waters.

"Not yet," Cohen replied. "But they'll be here."

Although nominally a freshwater animal, the zebra mussel nonetheless can tolerate salinities as high as four parts per thousand, enabling it to colonize estuaries and brackish waters. Like a saltwater mussel, it can attach itself to hard substrates; it made its new presence in the United States known through its tendency to carpet underwater surfaces with itself. A car pulled from Lake Erie in 1989 after eight months underwater was three inches thick in zebra mussels. Over a six-month period, workers at Detroit Edison watched as the density of zebra mussels lining the plant's intake canals jumped from one thousand per square meter to nearly three-quarters of a million per square meter. Two years later, when one of the company's twenty-foot-deep canals—by now 75 percent occluded—was blasted clean with a high-pressure hose, thirty tons of zebra mussels washed out. Navigational buoys have sunk from the gathered weight of zebra mussels. In December 1989 the town of Monroe, Michigan, was brought to a halt for two days when zebra mussels clogged the only intake pipe to the water-treatment plant. Between 1989 and 2003, the estimated cost to power companies alone of controlling the zebra mussel was slightly more than one billion dollars. Today the control of zebra mussels is an industry unto itself, a thriving marketplace of competing anti-mussel paints and sprays, filtration systems, de-cloggers, and allegedly surefire heat treatments—one set of opportunists met squarely by another.

The zebra mussel has struck ecological communities with equal force and speed. So exclusively do the animals congregate, so thoroughly do they colonize, crowd in, overlap, and overtake every available underwater inch, there is little room left for anything else. They are a monopoly of themselves. Native freshwater bivalves are the species most immediately threatened; even before the arrival of the zebra mussel, nearly three-fourths of the freshwater mussel species in the United States had been classified as rare or imperiled. Since 1988, nineteen species of native freshwater clams have disappeared from Lake St. Clair. The assault is sometimes direct: among the countless surfaces that zebra mussels can adhere to are the shells of native unionid clams. So many zebra mussels may attach to a clam—one scientist counted roughly ten thousand zebra mussels on one five-inch-long unionid—that the clam cannot move, nor burrow, nor open wide enough to feed or reproduce. The zebra mussel also inflicts starvation on its neighbors. Like other bivalve mollusks, the zebra mussel is a filter feeder: with a siphon, it pulls water into its mantle, then passes it through a set of small gills that screen and collect the edible particulates. It is a water-treatment plant in which the algal life, not the water, is the prized commodity. Tens of thousands of mussels (or more) per square meter, tens of thousands of square meters in a lake, equals hundreds of billions of tiny straws sucking, filtrating, consuming, removing every floating, edible mote from even the largest body of water. Weekly, the zebra mussels in Lake Erie filter a volume of water equal to the volume of the lake's western basin. In one study, researchers observed that the amount of chlorophyll in a *Dreissena*-infested lake dropped by half in a six-month period; the effect is much the same as if the Great Plains was suddenly shorn of half its grass. Scientists in Michigan recently discovered that Diporeia, a shrimplike crustacean that serves as the food base for large fish like salmon and trout, and which eats—or would eat—the same algae eaten by the zebra mussel, is rapidly disappearing from the Great Lakes.

"Ecologically, the zebra mussel could be bigger here than the Asian clam," Cohen said. He was sitting with an old car tire he had pulled from the water alongside the dock; a length of rope connected the tire to a piling, where it served as a yacht bumper. Cohen's hand was deep in the marine muck that had settled in the tire's well. "With the Asian clam, the concern is how much phytoplankton it's feeding out of the water. It's abundant; it lives in a part of the Bay that is lower in salinity but not fresh. The zebra mussel is freshwater to slightly salty. Their range would be nicely complementary. Both have the potential to be disastrous. Together they could really hog the food resources. Potentially. We just don't know until it happens."

Once, Carlton was invited to write a chapter for a particularly hefty book about the biology of zebra mussels. He opens by noting that freshwater and marine invertebrates "possess a remarkable variety of dispersal mechanisms"; then he goes on to illustrate that zebra mussels are particularly gifted in regard to motion and dispersion. Begin with the eggs. A single female zebra mussel may produce a million eggs in one year—eggs that may ride the currents for hundreds of miles and can survive out of water for several days. The larval phase can last anywhere from six to thirty days, if conditions are right, during which time the nascent zebra mussels drift as plankton, floating free with currents, down streams and rivers, through canals—through, say, all of western Europe. At last the time comes for a young veliger to find a suitable surface and settle. Only

3 percent of zebra mussels reach this juncture. Still, the end figure is considerable: thirty thousand adult zebra mussels from every million eggs, each year, per zebra-mussel mother. A young zebra mussel may pick up and move if its substrate does not suit it. Once firmly and permanently affixed, the adult may continue to expand its territory if by chance it has attached itself to a surface that is itself moving: driftwood, detached floats and buoys, plastic garbage, mats of algae, birds, amphibians, turtles, the wet fur of muskrats, beavers, even moose. Zebra mussels are known to adhere to live crayfish, which alone do not journey far; as bait, however, crayfish may travel hundreds of miles. Live mussels can be defecated or regurgitated by animals, although, as Carlton notes, the odds of successful egress diminish rapidly with time ingested.

If, when attaching itself permanently to an underwater surface, a zebra mussel wanted to maximize the likelihood that its chosen surface would soon relocate to a new body of water some distance away, the best odds would lie with a boat. Zebra mussels have been found on and in virtually every surface that boats possess: hulls, intake pipes, outtake pipes, outboard and inboard motors—on the hydraulic cylinders, trolling plates, prop guards, and transducers; on anchors, hawse pipes, rudders, propellers, shafts, and centerboards. Any recreational craft—yachts, dories, sailboats, canoes, rubber dinghies—will suffice, or any barge or workboat, or the small trailers that transport boats from one lake or river to another, or even the pontoons of catamarans and seaplanes. Not to be overlooked are the things that many boats carry from lake to river or from pond to stream: fish cages, buckets, bait and bait-bucket water, fouled tackle, nets, traps, floats, marker buoys—even, though less ideal, Carlton says, scuba gear, Windsurfers, water skis, and underwater cameras. It is mainly on surfaces such as these that the zebra mussel has managed to reach waters far removed from the Great Lakes: ponds in Vermont, reservoirs in Iowa, streams in Oklahoma.

"A few years ago I predicted that we'd have them here within ten years," Cohen said. "It's been found a half dozen times at the California border already, mostly in the piping of boat engines. I'm sure there must be as many that get through. I have no idea whether mussels attached to a boat can cause an introduction. Do you need a whole boatload?"

The zebra mussel's original incursion into the Great Lakes, however, was almost certainly mediated by ballast water—in a ship either from Europe or from the Black Sea, sometime in 1985 or 1986. It would have stood an excellent chance of surviving the crossing—either in its adult phase, attached to the interior walls of the tank, or as microscopic plankton floating in the ballast water itself. As Carlton himself discovered, ballast tanks regularly transport live organisms across wide oceans and do so in sufficient number to ensure that at least a few creatures will meet, mate, and reproduce in the receiving harbor. "It had never really hit the aquatic environment in this country where the interface with ballast water was so clear. We'd been waiting for the big disaster."

In 1990 and again in subsequent years, Carlton was called before congressional subcommittees to testify on the subject of ballast water and aquatic invasions. That year, Congress passed Public Law 101-646, the Non-indigenous Aquatic Nuisance Prevention and Control Act. Under the law, "aquatic nuisance species" are defined as any "nonindigenous species that threaten the diversity or abundance of native species or the ecological stability of infested waters, or commercial, agricultural, aquacultural or recreational activities dependent on such waters." The zebra mussel was Exhibit One. In an effort to slow the incoming tide of ballast-water invasions, the law also issued voluntary guidelines for all ships bearing ballast water into the Great Lakes. Today, before entering, all ships must first empty their ballast tanks (typically filled with low-salinity water from their home estuary) in the open sea, and refill them with high-salinity seawater. Presumably, the infusion of high-salinity seawater would kill or flush out whatever estuarine organisms remain in the tanks. (Any high-salinity organisms that might be drawn into the tank in the process would not likely survive their release into low-salinity port waters.) Since 1993 the laws have been mandatory for the Great Lakes. The U.S. Coast Guard regularly checks the salinity of the ballast water of arriving ships for evidence of an open-ocean exchange.

For marine biologists and policy makers alike, the Great Lakes have served as a kind of pilot program for the nationwide regulation of ballast water. The challenge lies in scaling up the program to match the daunting scale of the national shipping trade. In 1995, under a mandate from the Non-indigenous Aquatic Nuisance Prevention and Control Act, the U.S. Department of Transportation published a two-hundred-page report that quantified in great detail the role of shipping in the transfer of biological invaders: the number of ships entering U.S. ports annually; their origins and destinations; how much ballast water is carried; which ports are busiest. The report's primary author was Carlton; in assembling it, he gained an intimate appreciation for the difference between Great Lakes shipping and the overall enterprise.

"We're talking about a huge number of ships compared with the Great Lakes," Carlton said. He had resumed the helm of the minivan and was again navigating the Bay Area highways, with Cohen in the passenger seat consulting the map. Some twelve hundred foreign ships enter the Chesapeake Bay each year. The San Francisco Bay sees close to a thousand ships. The Great Lakes see maybe five hundred foreign ships a year. "And the Great Lakes are closed four months out of the year, from December first to March thirty-first. So it's a much more focused situation. The question is what to do about the rest of the U.S."

Carlton's acquaintance with the complexities of the shipping industry has granted him a finer understanding of the apparent trends and patterns in the emergence of new aquatic invaders. One question he commonly confronts is, Why did the zebra mussel invade in the 1980s and not, say, in the 1960s, shortly after the 1959 opening of the St. Lawrence Seaway? For that matter, why do other marine and freshwater invasions occur when they do, neither sooner nor later? The European sea squirt *Ascidiella aspersa,* a common fouling organism on the hulls of ships, colonized New England waters only in the 1980s, despite four centuries of regular ship trade before that time. If a man-made transfer corridor—a canal, a seaway, a shipping route—has been active for a hundred years, why does organism X not invade until year one hundred and one? If ships have been moving species around the world for centuries, why aren't the world's marine species everywhere already?

The reasons are myriad, Carlton says, and so expose the problematic nature of ecological prediction. In general, the odds of an organism successfully invading a new habitat depend on how it is introduced and what greets it upon arrival; those two factors can cancel or combine in any number of ways. Consider an organism traveling from Port A to Port B. The environmental conditions of Port B could change in such a way—cleaner waters, an influx of certain nutrients, or higher or lower salinity—that improves the arriving organism's chances of surviving and reproducing. Suddenly there are more Styrofoam floats in the waters of Port B, making a new habitat available for the burrowing isopod *Sphaeroma quoyanum.* Or the donor environment, Port A, might change in a way that increases the abundance of an organism, thereby increasing its odds of encountering a ship bound for Port B. Carlton offers the example of the Japanese clam *Theora lubrica*, which appeared in San Francisco Bay in the 1980s. Only a few years earlier, a cleanup of the Inland Sea of Japan—a significant donor of ballast water to San Francisco Bay—had significantly increased the number of clams living there.

Or perhaps the environment changes not at all, and instead the vicissitudes of commerce alter the flow of ballast water between one port and another. The Asian clam *Potamocorbula amurensis* did not appear in San Francisco Bay until the 1980s, in part because before that time, direct international trade between China and North America was prohibited. Its success on arrival may have been abetted by climatic factors: a period of heavy rain and flooding around then may have lowered the salinity of San Francisco Bay sufficiently to improve the claims' survival rate. It should be noted that ballast water travels largely against the main currents of trade. Ballast is intended to steady an empty ship—that is, a ship traveling somewhere to pick up a load of something commercially valuable. Therefore, ballast water, and the aquatic life therein, tends to move from the harbors of goods importers to the harbors of goods exporters. The exotic creatures in one's port waters, then, provide some historical insight into what humans have come from afar to acquire. In the late 1980s, marine scientists in Tasmania noted with alarm the sudden arrival of a predatory sea star and a pernicious, fast-growing seaweed, both native to Japan. Tasmania's main export is woodchips from the island's dwindling forests; the principal buyers are Korea and Japan, which send empty, ballasted ships to collect it. Since the 1960s, those ships have become ever faster and larger, bearing more ballast water—and with it, more kinds of organisms and more individuals per species—per load, all in all granting the organisms a higher likelihood of surviving their voyage and of mating and reproducing upon release. And every harbor that receives and fosters an exotic organism becomes a potential donor of it. In the 1980s the East Coast saw the arrival of two Asian marine species, the sea squirt *Styela clava* and the green algae *Codium fragile tomentosoides*; they arrived not from Asia, however, but from western Europe, which had acquired them some years earlier. Within a few short years of receiving the zebra mussel, the Great Lakes is now an inadvertent exporter of it.

In a recent paper with Gregory Ruiz, a former marine ecology postdoc with Carlton and now the director of an invasions lab at the Smithsonian Environmental Research Center in Maryland, Carlton boiled down the various factors governing invasion success into a single mathematical equation. Its apparent precision disguises a great deal of ineffability:

$$I = \sum_{i=1}^{n} ((Ps_i) (R_i) (B_i))$$

What it states generally, in vague concision, is that invasion is a function of three variables: the incoming propagule supply (Ps), the state of the recipient ecosystem (R, which stands for "resistance"), and the bias of the data (B). The precise contribution of each variable varies, depending on the organism and the situation. The easiest variable to measure, and the one that seems to carry the most force, is propagule supply: the number of organisms that arrive and the frequency with which they do so—in ballast water, on ship hulls, in airplane wheel wells, or by whatever other means. The biggest unknown, however, and the variable most likely to throw off everything else, is B, the bias of the data. The equation can proceed accurately only based on what scientists know for certain. The facts they don't know—and that they don't know that they don't know, the unknown unknowns—are a large part of why ecological invasions are nearly impossible to predict in advance and probably will remain so. In retrospect, Carlton says, the opening of trade with China in the 1980s might have suggested that the Asian clam *Potamocorbula amurensis* would soon invade San Francisco Bay—except that, owing to the trade restrictions, most of the Asian biota had never been seen or studied outside of native habitats, so biologists had no reason to even consider *Potamocorbula* on the list of potential colonizers. So it goes with most invaders still.

"In reality, we do not know why the zebra mussel, or any other recent invasion, was successfully introduced when it was, and not earlier," Carlton wrote in the Department of Transportation shipping study. "Similarly, we cannot explain why many species have not yet been introduced into North America." Now, in the van, he said, "The existence of a transport mechanism doesn't tell you that everything that could have been introduced would have been introduced by now. That's a very common argument: 'Ballast water has been dumped in here for decades—why don't we have everything already? Why didn't the zebra mussel enter the Great Lakes in the fifties, or sixties, or seventies?' The answer basically is we don't know. The likelihood is that there is no one answer. It may be a synergism of things that have happened around the world. The most popular hypothesis is, 'Well, you should have seen the Great Lakes back then.' You know, light a match, throw it on Lake Erie. For some people, the invasion of the zebra mussel is what you get when you clean up the environment."

If the thrust and impact of Carlton's globe-spanning and interdisciplinary career were to be summarized in a single word, that word would be *cryptogenic.* Although Carlton did not invent the word—in medicine it indicates a problem without a clear cause—he is first to apply it to the realm of marine biogeography.

In 1979, with an exhaustive survey of the introduced marine invertebrates of the Pacific coast to his credit, Carlton took a postdoctoral position at the Woods Hole Oceanographic Institution. There he proposed to conduct an exhaustive survey of the introduced marine invertebrates of the Atlantic coast from Canada to Florida. "I figured, 'This is easy, I'll knock it off in a year.'" It wasn't, and he didn't. "Suddenly I realized the depths to which the history of biogeography was not known. It was here that I faced the mystery of shipping traffic."

On the Pacific coast, where the history of shipping dates back to Spanish explorations in the 1500s, Carlton had tallied three hundred introduced organisms. His historical review of the marine-science literature of the East Coast, however, turned up references to barely a handful of introduced species—despite the fact that ships have visited the Atlantic coast in abundance since the seventeenth century and in smaller numbers for centuries before that. Carlton was baffled by the discrepancy and suspicious of some of the allegedly native candidates. Consider *Sphaeroma terebrans*, a saltwater isopod that burrows into the root tips of living red mangroves. Although it is considered indigenous to Florida, it also resides along distant coasts of South America and in the Indian Ocean. That is a large and unusually disjointed range for a supposedly local species. Was this organism, and the oddly cosmopolitan organisms like it, truly native, as the East Coast scientific literature insisted? Or were they in fact elder exotics, introduced before scientists arrived to study them?

"All my senior colleagues were saying, 'Oh, no, it's all natural.' If it's described as being seen in Long Island Sound in the 1820s, it's assumed to be natural. I'm the guy from California; this was not my turf. 'Cryptogenic' was my way of fessing up to what I don't know." In Carlton's usage, a cryptogenic species is a species of uncertain background, neither indisputably indigenous nor nonindigenous, not clearly from here or there. Between opposing territories—the purely natural landscape and the landscape modified by human handling—Carlton had wandered into a vast gray no-man's-land. "There had been a dichotomy of life, the natural versus the human-mediated," he says. "In New England, I found the trichotomy of life."

In 1956, in a book titled *Man's Role in Changing the Face of the Earth,* the natural historian Marston Bates contributed an essay about biological invasion in which he wrote, "It becomes difficult to draw a line marking off the human habitat, and there is every degree between human dominance on Manhattan Island and human insignificance in the forest of some remote tributary of the upper Amazon. Yet even in the remote forest we may come across a mango tree, the only trace of a Jesuit mission abandoned a century and a half ago; and there are always the small, shifting clearings of the Indians. As biologists, we are apt to deplore this, to brush it off, to try to concentrate on the study of nature as it might be if man were not messing it up. The realization that, in trying to study the effect of man in dispersing other organisms, I was really studying one aspect of the human habitat came as a surprise to me. But, with the realization clear in my mind, I wonder why we do not put more biological effort directly into the study of this pervasive human habitat."

Carlton likewise was surprised by the depth of his own discovery. In 1992, in a paper titled "Blue Immigrants: The Marine Biology of Maritime History," he noted, "So old and numerous were sailing voyages between Europe and North America—beginning in the sixteenth century, but extending back to tenth-century Norse explorations—and so recent are our first biological investigations, few being available before the mid-nineteenth century, that it is difficult to determine for many species whether they occurred in America before European contact." Sitting on a sofa in Mystic, with a plastic globe on a stand in front of him and his finger drifting above its currents, he said, "When biologists appear, around 1800, they take the world to be a natural place. We're seduced into this all the time. If a species occurs from Nova Scotia to Florida, or from Norway to Spain, or from Alaska to Mexico, we assume it's been there since time immemorial. But you cannot use the breadth of occurance as evidence of where a species used to be. The colonists arrived with the maps—and they were preceded by the explorers who made the maps. You see the footprints of human travel before you see the roads that follow the footprints."

Where does the natural world end and the human world begin? And what exactly lives in between? Carlton estimates, conservatively, that between the years 1500 and 1800, three marine species a year were successfully, if inadvertently, introduced to new parts of the world on the hulls of sailing ships. Three species a year for three hundred years: "That's about nine hundred species. I refer to them as the missing one thousand." One thousand species that by standard definition are nonnative yet are marked in the scientific record books, mistakenly, as indigenous. Carlton has fixed a number to them; so far, however, their actual identities are anyone's guess. Like former criminals in the federal witness-protection program, the missing one thousand reside among us as historical ciphers: blank-faced and quietly busy, familiar at a glance, perhaps, yet entirely opaque on close inspection. "I'm not completely pessimistic about the prospect of sorting it out. It will require the work of systematists, historians, geographers. I don't think those missing one thousand are entirely lost to us."

Other marine biologists have also been surprised by Carlton's discovery, not all of them pleasantly. "Many of my senior colleagues take certain aspects of species to be natural—'It's amphi-Atlantic, don't you know.' Many a finger has been wagged in my face. They give all sorts of explanations that sound plausible at first: plate tectonics, ancient seaways and land-island corridors, glaciation, the opening and closing of various barriers. And of course natural dispersal, despite the fact that these cosmopolitan species don't have larval forms that would permit oceanic crossings." A particular subset of skepticism has been reserved for Carlton's notions about the inadvertent transfer of microscopic algae—single-celled diatoms and dinoflagellates—around the world in the ballast water of ships. There are countless species of such organisms, tens of thousands at the very least, of which only a relative handful have been identified and named by taxonomists. Most are innocuous from a human point of view, but a few, given the right water conditions, are known to reproduce in sudden, massive numbers to create toxic, miles-wide blooms in the ocean. Shellfish that consume large numbers of such algae are themselves unsafe to eat. A rise in cases of paralytic shellfish poisoning, which can cause severe respiratory distress or even death in a person who eats a toxic mollusk, is thought to be linked to the occurrence of harmful algae blooms. The number of "red tides" around the world has jumped markedly in the past three decades; Carlton believes ballast water is partly to blame.

"From the 1970s to the 1990s, twenty to twenty-five years when there are innumerable demonstrable invasions by ballast water all over the world—crabs, mollusks, worms, snails, jellies, and everything else—is the same period of time when red tides bloom all over the world. The same period. Often you find dinoflagellate species that were never before seen in these areas. Yet dinoflagellate people say these are not ballast-water introductions, that these species have always been there, but for some reason they are blooming now. This has been the subject of seven international conferences in twenty years. There are thousands of pages written as to why these red tides are blossoming: eutrophication, global climate change—endless hypotheses. If it was a crab instead of a dinoflagellate, they'd say it was introduced.

"The hypothesis that in fact a great many of these are introductions by ballast water has not been seriously examined by these guys. It's difficult for me to understand. In San Francisco Bay you have more than two hundred species of introduced animals and larger plants. But when you ask phytoplankton people what's introduced—in a very large, diverse flora of diatoms and dinoflagellates—they don't regard a single species as introduced. That's simply impossible. Diatoms are one of the most common species in the world in ballast water. Given all the ballast water that's dumped from all over the world, you'd think we'd have European diatoms, Australian diatoms, just as we do all the other invertebrates. And yet they can't recognize them. Of course, I'm sympathetic to the taxonomy: they're extremely hard to identify." In their published survey of the Bay, Carlton and Cohen found some two dozen diatom species of sufficiently murky nativity to earn the label "cryptogenic." The response from phytoplankton scientists is expected to be unfavorable. "All we want to say is, 'We call them cryptogenic, which means we don't know that they're introduced and you don't know they're native.' We think their response is going to be, 'Yes, we do.' We're waiting for that. We want to say, 'How do you know?' The default in biogeography is generally, if you don't know, it's native. But in fact, if you don't know, you should say you don't know."

In effect, Carlton threatens to turn classical biogeography on its head. The traditional overlook takes a sort of innocent-until-proven-guilty view of the seascape. The distribution of most organisms along the continental margins is assumed to be natural, unless specific historical evidence of human-induced change argues otherwise. Carlton would reverse the burden of proof. In a paper titled "Man's Role in Changing the Face of the Oceans" he writes, "The overwhelming nature of human-mediated dispersal mechanisms of marine organisms . . . may be leading us to consider, not how we know a species is introduced, but how we know a species is native." And not only individual species, but the appearance and contents of entire marine ecosystems may be historically suspect. Consider again *Sphaeroma terebrans*, the saltwater pill bug allegedly native to Florida. Carlton now believes that the organism in fact is native to the Indian Ocean; it probably arrived in Florida in the 1870s as a fouling or boring organism on visiting ships. The creature lives by boring holes in the propagating root tips of red mangroves from Florida to South America, effectively halting the seaward advance of these mangrove forests. For more than a century, beginning before the first naturalists arrived and documented the purportedly native flora and fauna, *Sphaeroma terebrans* has single-handedly reshaped a large stretch of the tropical Atlantic coast. Absent this organism, what shape would the "natural" coastline take? Human-mediated invasions have been so fundamental, pervasive, long-standing, and overlooked, Carlton writes, "that we may never fully know what the biota of the continental margins looked like before ships and before the movements of commercial fishery products."

On land, a naturalist can navigate the temporal landscape with some confidence, comforted by the relative solidity of historical knowledge. There, it is possible to compare the appearance of nature before the arrival of humans with its appearance today, and then to chart a path toward how to conserve it for tomorrow. In contrast, the marine naturalist is adrift in time, bearingless. This discomfiting fact, Carlton acknowledges, has profound implications for marine conservation. In 1972 the federal government established a series of protected estuaries—the National Estuarine Research Reserve System—"to serve as natural field laboratories in which to study and gather data . . . so that scientists will be able to study the naturally functioning system." Twenty-six such reserves now exist throughout twenty states and Puerto Rico, spanning more than a million acres of estuarine waters, marshes, shoreline, and adjacent uplands. What exactly are these reserves preserving? How natural, how native, are they?

Carlton can't help but wonder. Some years ago he conducted a biological survey of the South Slough reserve in Coos Bay, Oregon; it is one of four estuarine sanctuaries on the Pacific coast, and it lies not far from where he conducted his first ballast-water experiment. In South Slough, he counted at least thirty-two species of nonnative marine organisms, including dense stretches of Japanese eelgrass, which in turn hold several species of introduced worms in great abundance; vast stretches of the Atlantic sponge *Halichondria bowerbanki*; and the Japanese orange-striped anemone *Diadomene lineata*. With few exceptions, all the nonnative species Carlton found had probably been in place before the South Slough reserve was established in 1974; in some cases, they arrived long before. And that tally counts only those introduced species that are known to and, after concerted effort, could be identified by a certain expert in the marine invertebrates of the Pacific Northwest. It does not include the epistemologically unknowable species—the members of the "missing one thousand," should any be in residence.

How these introduced species—the knowns, the unknowns, and the unknowables—may have altered the biota and environment of South Slough before scientists first arrived to label it is itself unknown and perhaps unknowable. However, one can safely assume that the other estuaries on the Pacific coast, and probably across the country, are equally "natural" in imagination only. "Few works in coastal marine ecology, a discipline that began in the late 1940s, offer any robust evidence that the systems studied are completely natural," Carlton has written. "Restoration projects that seek to restore a bay, an estuary, or a marsh to the way it looked when the proponents were young—as if this temporal target was by default the aboriginal world—seek a world that was already a barely detectable shadow of its former self."

Where this insight leaves the marine scientist is uncertain. How does one proceed when the alleged baseline for a naturally functioning system—the "natural field laboratories" against which one might compare the traits of seminatural or wholly man-made ecological systems, to see how they differ—turns out to be an unreliable representative? How do you learn from that?

Late one afternoon, the expedition crew found itself at a marina in Redwood City, a prosperous suburb a dozen miles south of San Francisco. It would be the last stop of the day before heading back into the city for a long night in the lab with the microscopes and the taxonomic monographs. Carlton was on his stomach, reaching into the water for a mass of orange sponge—*Clathria prolifera,* originally an Atlantic species—that was visible on one of the floats supporting the dock beneath him. The evening was a lovely one, featuring a slight breeze and a gibbous moon rising over the yacht masts. The marina was situated directly under the landing path of San Francisco International. Every two minutes a massive jetliner—Singapore Air, American, United—swept low overhead, its landing gear extended like talons. The sponge was just beyond Carlton's grasp.

"These newer docks!" He laughed and sat up. "That's a beautiful sponge. It's called the red-beard sponge on the East Coast. None of the sponges we've seen so far are native." There was nothing to suggest that any of the Bay's native sponge species were less abundant as a result of this sponge's presence, Carlton said. If *Clathria* was displacing anything, it was displacing things like mussels and barnacles, most of which were likewise nonnative. Carlton peered down through the water at the assemblage of exotics. It had a discernible structure: a base of mussels, *Mitilus galloprovincialis,* that attached directly to the float. Stuck atop them was a mixed forest of yellow and orange introduced sponges, through which introduced isopods undoubtedly trundled. "It's a very handsome sponge fauna here. Sponge City! The interesting thing is, all these species evolved in different parts of the world: Japan, Australia, the East Coast. And here they are living together."

That such far-flung species can so readily congregate and cohabitate is an abiding mystery in ecology. Formally, a species is a grouping of organisms that are genetically (and thus anatomically) similar—similar enough that they can interbreed and so pass on their genetic similarities. The stripe pattern of individual zebras may vary from one to the next, but zebras as a species are nonetheless distinct. In effect, a species is an entity in time. Through natural selection, it has developed a set of traits—stripes, say, which disguise the zebra from predators on the open veld—that enable it to persist for generations, through centuries, millennia, even eons. It is a ship in a bottle on the temporal sea.

The same axiom is commonly thought to apply to local collections, or communities, of species as well. Individual species are adapted to both physical conditions (average temperature, amount of sunlight, and available nutrients) and the presence of other species. The evolutionary relationships and dependencies that arise over time sometimes can be highly specific, like the deep-throated Hawaiian blossom that is pollinated only by certain species of long-billed Hawaiian honeycreepers. Together, these interdependencies are thought to confer a degree of collective stability, which enables individual species—and their ties to other species—to persist. This network of ties generates barriers to invading species and so permits the entire community to persist intact through time. According to this logic, mature communities, comprised of species that have long evolved together, are fundamentally different—more tightly integrated, more stable, more productive—than communities that are tossed together quickly and at random. Allegedly, Carlton said, evolutionary time matters. And if time matters, its varying products—mature communities of coevolved species versus recent, tossed-together collages of invaders—should be readily distinguishable.

"If I said, 'Pile all these things together,' could you tell me whether they were introduced or not, by some way in which they have inserted themselves into the system and interact with other organisms? It's an interesting way to look at some assumptions of evolution and how communities come about. I mean, there are some places in the Bay where virtually everything is introduced. They come from ten different places in the world, so they're not coevolved. You'd expect that these communities would somehow be different from communities of species that have been together for thousands, or tens of thousands, or hundreds of thousands of years—or even older systems. That's what community ecology is all about. Because when we go to communities and look at interactions, we mostly assume that many of the species have been together since time zero."

So, he added, what should a viewer make of ecological metropolises like Sponge City? If true differences exist between untouched, natural ecosystems and human-modulated communities, how is it that for nearly a century now, marine biologists have studied the dynamics of allegedly "natural" systems—in South Slough, on the New England coast, in the Florida mangroves, around the world—and never realized their mistake? "I guess one of the things that this mélange of exotic species says to me is that since a lot of our colleagues are surprised when you look at a community like this—it does not speak to you immediately: 'Oh my gosh, what an artificial assemblage!'—it makes me think that one can sew together and assemble a community of species quickly, and it will fool you. So what does time matter?"

In the late 1980s James Drake, an ecologist at the University of Tennessee in Knoxville, conducted an elegant laboratory experiment to explore exactly that question. He began with a group of fifteen small species common to ponds and lakes: four species of algae (the producers), seven species of tiny crustaceans (which would prey on the producers), plus four species of bacteria and protozoa. One by one, over a period of weeks, he dropped the species into five-gallon aquariums of enriched water and watched—for as long as a year, in some cases—as they formed what he called microecosysems. He repeated the exercise ten times, each time altering the sequence in which the various species "invaded" the microecosystems already taking shape. According to classical ecology, the assembly of ecosystems follows a somewhat deterministic course: if one could turn back the clock in a place like Hawaii and rerun the sequence of ecological invasions that occurred over geological time—if one could insert the same species all over again at the same time intervals as before—one would end up with pretty much the same groups of plants and animals, in much the same relative proportions, that exist there today (or did, before humans arrived). Drake reasoned similarly. If it is true that the structure of ecological communities is shaped largely by environmental factors (which remained constant across all the iterations of his experiment) and the evolved traits of individual species (which shape the interactions between species and, presumably, the kinds of communities that can arise), then he should see some consistent pattern between the order in which species arrive and the kinds of communities that take shape.

That was not at all what he found. Indeed, his results were sufficiently random as to be profoundly enlightening. Drop by taxonomic drop, Drake managed to produce numerous persistent, apparently stable communities, but never the same one twice. Occasionally two roughly similar ensembles of species formed through two entirely different sequences of invasion. More often, a single sequence of invasions, when repeated, gave rise to entirely different assemblages. Having watched one community take shape, Drake could not re-create it, even by repeating the recipe exactly. (This experiment, he recalls today, "nearly killed me with endless hours of microscope work.") One assemblage might show itself to be more resistant than another to the invasion of a particular species. Then again the same assemblage, formed through a different sequence, might be less (or more) resistant to invasion by exactly the same species.

The evolutionary factors typically thought to determine how communities form—the inherent abilities of individual species to compete, catch prey, or evade predators—played little or no part, Drake concluded. "Specific mechanisms (e.g., predation, competition, and emergent properties such as food-chain length) are frequently insufficient to characterize the patterns seen in ecological communities." What mattered were history and chance: which species showed up first, or tenth, or last, and the abundance and groupings of the other species already on hand to greet them. History didn't apply forward. From a given sequence of invasions, one could not predict in advance which assemblage would result. Whatever "rules" could be said to have governed the assembly of any particular ecological community are inextricably bound up in—and are individual to—the chance events in that community's past.

Classical ecology proposes that individual species play certain ecological roles—determined over time by natural selection—that they would fulfill regardless of the ecosystem they end up in, and which determine the overall shape and membership of that community. Drake's experiment instead supports what he calls an "alternative states theory": the same pool of species will give rise to a variety of ecosystems, even if the sequence of their introduction is identical. Drake says, "It's like taking apart a Volkswagen and discovering you can put it back together as a Mercedes." A writer is drawn to a literary analogy. Give twelve monkeys a turn at a typewriter, a standard keyboard with twenty-six letters, and they may well churn out *Hamlet*—once. They may also produce, once and once only, *Walden Pond, The Lost World, War of the Worlds, Around the World in Eighty Days, À la recherche du temps perdu,* and the collected exploits of Tintin. They will also fail repeatedly: non sequiturs, misspellings, fragments, gibberish.

James Drake is now the editor of *Biological Invasions,* a quarterly scientific journal that Carlton started in 1998 and edited through its first five years. I called Drake in Tennessee one afternoon; he had just discovered two slugs on the stairway outside and had come indoors to try to figure out if they were an introduced species. (They were—both members of *Testacella haliotidae,* an attractive invader from Europe.) I posed Carlton's question: Does time matter? On the one hand, Drake said, not really. The traits of individual species, the adaptions they have gained through evolutionary time that permit them to to coexist with their neighbors, are amazingly irrelevant in explaining why an ecological community is assembled one way and not another. "The dynamics are so rich. Competition, predation—those have little to do with it." Older communities, in which the member species have coevolved for relatively long periods undisturbed, are neither more nor less vulnerable to invasion than younger communities. The very notion of resistance is tautological, he added. "Systems that have been around a long time are, de facto, systems that haven't seen a lot of invasions. It's a bit of a circular loop."

On the other hand, he said, timing—the chronological sequence in which ecological communities form—is everything. "A lot of what we see is context-derivative. It has a serious historical dimension." Although studies of competition, predation, and the other dynamics that occur between species are genuinely useful for understanding how the current structure of an existing community is maintained, they do not shed much light on how that structure originally came to be. The X-ray analysis of a door latch may reveal how a car door opens and closes, but it won't explain why the car parts add up to a Mercedes rather than a VW. For that, you need to review the assembly process.

Likewise, to understand why an ecological community is the way it is, a scientist must both understand it in the present and retrace the series of largely unrepeatable historical events that led to its formation. For many marine communities, Carlton now finds, that history is precisely what is lost, perhaps irrevocably. The wake of human maritime history washed over marine biological history so thoroughly, and so long ago, that it is impossible now to envision what the seascape might have looked like without us. Unanchored by a definitive past, a marine scientist floats in the eternal present, like a sentence on the printed page. You can read the finished line, but you can never glimpse the crafting hand—its insertions, erasures, second thoughts—that honed it finally to a single word: *cryptogenic.*

18

Land is a static medium. From the point of view of all but the one species that has invented time-lapse photography, retraced Earth's tectonic history across the eons, and, from the Moon, watched the home planet spin on its axis, land does not move. Terra firma is firm. The sea is the opposite: it is a medium in constant motion, its waters driven horizontally, vertically, and in vast, basin-spanning gyres by wind, sun, heat, salt, gravity. And the flow of the sea is grasped, however faintly, by every animal in it. If you are not drifting in the current, you are waiting for something to drift by. To live on land—unless you are a plant and have the luxury of dining exclusively on sunlight—you cannot remain at rest; you must move. In the sea, the movement happens for you; indeed, it is all you can do to stand still.

No—to stand still, you must be a mussel or, better, a barnacle: crusty stalwart, paragon of persistence. There are innumerable species of barnacles belonging to two general categories. Some attach themselves by means of slender stalks and are known, loosely, as goose barnacles. The others, called acorn barnacles, clamp directly to their chosen surface. Some goose and acorn barnacles adhere to wood or stone, docks or rocks. Others live exclusively on the bodies of whales and survive years of ocean travel by setting deep anchors in the cetacean's thick hide. Late one night in New York, entangled in *Moby Dick,* I came across a mention of these peripatetic barnacles. Melville describes a particularly accurate image of a whale hunt by a French artist he calls Garnery (probably Louis Garneray): "In the second engraving, the boat is in the act of drawing alongside the barnacled flank of a large running Right Whale, that rolls his black weedy bulk in the sea like some mossy rock-slide from the Patagonian cliffs . . . Sea fowls are pecking at the small crabs, shell-fish, and other sea candies and macaroni, which the Right Whale sometimes carries on his pestilent back." I have seen photographs of whales encrusted with barnacles, of course, yet somehow I had forgotten that barnacles are actual living animals; nor had I stopped to imagine that a whale might carry a constellation of other creatures on its skin. Intrigued, I shot off an e-mail query to Carlton: What exactly lives on whales? Are whales a migratory platform for other marine organisms? Do creatures hop on and drop off as accidental tourists—settle onto the whale's skin in, say, Mexico and fall away in Majorca? Could whales, like ships, spread exotic species?

Carlton must have been working late, because his reply arrived only a few hours later, at six o'clock in the morning. "A number of whale species have a well-known associated epizoic fauna," he began. The most common living appendage of whales, he said, are a class of small crustaceans called cyamids or whale lice; these were most certainly the "small crabs" to which Melville referred. There are also numerous species of whale barnacles, belonging to two broad types: a sessile, or stationary, form that settles directly on and in the mammal's skin, and a stalked form that attaches to the sessile barnacles. By "shell-fish," Carlton said, Melville probably meant the sessile barnacles. He added, "The 'other sea candies and macaroni' may either be some reference to the masses of draping stalked barnacles, or a literary expansion." All of these species, Carlton went on, are uniquely coevolved with whales.

Although whale barnacles occasionally turn up on ships—a latter-day arrival on the seas—the converse does not occur: the types of barnacles, algae, sea stars, and other organisms that foul ships are not seen on whales. This is no surprise, Carlton wrote, as whales spend little time in the low-salinity harbors and estuaries favored by ship-fouling organisms. "It wouldn't surprise me if gray whales in the Baja lagoons occasionally pick up some inshore species (such as the ephemeral weedy green algae *Enteromorpha*)—but apparently these do not survive the subsequent ocean trips, since, as far as I know, no whales have been reported as 'greenish,' or draped with seaweed, etc." In sum, he said, whales and their migratory compatriots—sea turtles, sea snakes, dolphins, and other marine animals that often become encrusted with barnacles and smaller organisms—are not vectors for exotic species. However, Carlton hadn't noticed that passage from Melville before; he was eager to track down the Garneray engraving for his students. "It would make a great slide!"

Whales aside, the most familiar kind of barnacle is the acorn barnacle, typically seen on intertidal rocks, on dock pilings, and on the hulls of ships. The acorn barnacle lives in a calcareous house—eight plates of shell that it constructs around itself to ward off predators. When exposed to the air, this shell enables the animal to retain its moisture and salt balance. At low tide, the house is clamped tight, a sharp hazard to bare feet. At high tide, feathery cirri fan the water for food particles and retrieve whatever nutrients pass by: a field of eyelashes batting underwater. For a long time, the barnacle was assumed to be a mollusk, of the same inert order as scallops, clams, and mussels. In fact, the barnacle is a crustacean; its nearest cousins are shrimps, lobsters, and tiny krill. Picture the barnacle as a shrimp lying prone inside an armored igloo. The feathery fronds it extends into the water to capture food are neither tentacles nor antennae: they are the animal's feet, which have evolved into something akin to a sweep net. The barnacle is perhaps the only animal whose limbs enable not motion, but rather a life entirely devoid of it. Indeed, unlike every other crustacean, the adult barnacle is immobile. If it moves, it is only because it has permanently attached itself to a surface that moves. Yet the stationary phase is only the most visible and familiar phase in the barnacle's long life span. The barnacle is what is known as a broadcast spawner, the marine equivalent of a dandelion. During reproduction, the eggs of a barnacle develop and hatch within the safety of the adult's shell. (Fertilization requires a probing visit from the tubular penis of another stationary barnacle nearby). When the time is ripe, the armored door to the ocean opens, and the newly hatched barnacles, known at this early stage as nauplii, are released on the currents by the thousands. The children are sent to sea.

One afternoon in the laboratory in San Francisco, after several long moments peering into his microscope, Carlton looked up suddenly and exclaimed, "These *Balanus improvisus* have just released nauplii into my sample!"

Carlton was examining a sample of specimens the expedition had collected early that morning. He motioned for me to take a look. I gazed down upon mayhem: pale, vaguely crablike blobs, each one perhaps a millimeter in diameter, hopped in all directions. They were the newly hatched young of the barnacle *Balanus improvisus*, a common species on the East Coast and, increasingly, on the West Coast. The larvae represented the beginning of the mobile phase of the barnacle's life, the start of their moving on. "My *Balanus improvisus* are releasing nauplii!" Carlton said again. "They look like little, jerking . . . nauplii. I'm running out of adjectives."

A larva or a seed is a propagule, a tidy ship of genetic information built to traverse some amount of geographic space and found a new generation. In the halls of academia, those vessels are called graduate students and postdocs. Carlton has produced several thus far. "I've had five postdocs, all of whom are doing invasions now and have set up their own labs," he said at his microscope. "And even more graduate students." The transfer of knowledge marks a course shift of his own. Having launched the field of marine bioinvasions and ballast-water ecology, Carlton is slowly turning his attention to the horizon. Recently he has published several small scientific papers of which he is proud, on the subject of what he calls neo-extinctions: the recent, previously unacknowledged extinction of organisms, mostly invertebrates, in the sea. He is fascinated by the issue of extinction in the ocean.

"When we talk about extinctions in the ocean, we never talk about extinct fish, extinct invertebrates, extinct seaweeds. We just talk about whales and the older stuff—Steller's sea cow, the great auk—that people no longer have any memory of. We never talk about extinctions in the rest of the ocean. That's interesting to me. And the search for those extinctions is interesting to me. I've spent some thirty or more years looking at the additions to communities; now I'm interested in the deletions to those communities. It takes quite a while to dig out any evidence. On the other hand, it's a much less busy field than invasions, and I'd probably get to stay home a lot more. Nobody besides me has any active research interest in looking at invertebrate extinctions in the ocean. Of course, we could have said that about ballast water five or six years ago. Now there are dozens of people attending to it."

One extinction of particular concern to him is the looming extinction of marine invertebrate taxonomists. One day soon, Carlton will publish the fourth edition of *Light's Manual*, the indispensable handbook to the taxonomy of intertidal organisms of the West Coast. The manual will represent the collected insights of more than a hundred taxonomists: specialists in systematics—scientists employed by museums mostly—who know their narrow field of zoology so well that they can distinguish, say, the microscopic, multi-legged adult form of an Atlantic copepod from the microscopic, multi-legged adult form of a Pacific copepod. To no small degree, the current concern with biodiversity—how many of which species live where, and how quickly they may be going extinct—relies on the existence and involvement of taxonomists. Most of the biologists combing the rain forests and high seas for new species of organisms do not themselves have the expertise to identify said species on the fine scale, and instead must send them off to a taxonomist to study. Ultimately, an accurate measure of biodiversity is limited by the number of taxonomists on hand to measure it.

Alas, Carlton said, their number is diminishing; taxonomy is a dying art. To a non-taxonomist, the science of systematics is about as exciting a job track as stamp collecting. Universities have little interest in funding professorships in what they deem to be cataloging; the number of academic posts is dwindling. Consequently, the premier taxonomists are in their sixties, seventies, or eighties; few young taxonomists are rising to fill the ranks. Field biologists have fewer and fewer places to send their specimens for identification. "It's great to see this concern with biodiversity," Carlton said, "but I don't see it being matched by an increased level of training. Lots of us can train systematists. But to what end? Where are they getting jobs? Everybody says they need taxonomists: all the hands go up, let's put a big check in the 'positive' box. But when an assistant professor job opens up, are you going to hire a taxonomist? 'Ummm, er, ahhhh . . . we need an experimental ecologist . . . we need a molecular biologist.' The emperor has no clothes." Carlton finds himself at the diminishing end of this curve. With a mounting awareness of time, he has long sought a collaborator for the upcoming edition of the *Light and Smith Manual.* "I'd like to find a sparky young person in their twenties or thirties. Such is the state of taxonomy that I've not found a person of a younger generation whom I could seduce into being a coeditor. Only now do I understand what Ralph Smith was doing when he brought me onto this project twenty-five years ago. He was grooming me to take it over."

A barnacle obeys a circular life cycle that leads it from solid ground (as a new larval nauplius) out to sea and then back again, to settle as a calcareous, sedentary adult. Along the way, the larva passes through several intermediate morphological phases, each one remarkably adapted to the different water conditions the organism will encounter in its odyssey.

Consider a barnacle located in a coastal-plain estuary like San Francisco Bay, which is fed at one end by the Sacramento–San Joaquin River and opens at the other end into the sea. Freshwater runs from the river into the estuary and toward the mouth of the bay. Saline water is more dense than freshwater, so as the river empties into the estuary, its freshwaters flow out along the surface. Seawater moves in to take its place; being more dense, and thus heavier, it drifts toward the head of the estuary along the floor of the bay. The net result is a circulation of salinity: a fresh current above flows out and, as it mixes with seawater, gradually sinks; a saline current below flows in and, as it mixes with freshwater, gradually rises. The life cycle of *Balanus improvisus* has evolved to ride this gyre. A barnacle nauplius passes through six successive stages; each stage comes with added anatomical ballast, such that the nauplius rides successively deeper in the water column. As a result, the earliest nauplii are carried toward the sea while the elder ones are carried toward the estuary. By the time the nauplius has passed through its last molt and adulthood is imminent, it has returned more or less to the place where it began.

At this point the barnacle larva has entered into its seventh and final juvenile stage and is called a cyprid. A cyprid has a bivalve carapace; one might picture it as a tiny clam with legs, except a cyprid also possesses two compound eyes, like a fly. It begins to search for a permanent home. The cyprid drifts until it encounters a surface or substrate to its liking; probing with its antennae, the animal may take hours or days in choosing a suitable landing site. (Among other criteria, cyprids seem to prefer areas in which other barnacles have already settled.) Then, with the aid of a brown glue it exudes from adhesive glands in its antennae, the cyprid attaches itself headfirst to terra firma. Once rooted, the animal undergoes a profound metamorphosis: its eyes disappear; its swimming appendages change into feathery, food-gathering cirri; from its carapace it secretes the armored plates that will lock it down forever. As a caterpillar cocoons itself and is transformed into an ultra-mobile butterfly, the cyprid performs the reverse transformation, from motile to inert. Here the barnacle will remain for life: a shrimp doing a headstand, antennae to the ground, legs in ceaseless motion, gathering, consuming, digesting, broadcasting—until the ravages of time and salt, or a man with a pocketknife and a jar of formalin, dislocate it from its tenure.

19

Late one warm November afternoon I stood on the bridge of the SeaRiver *Benicia* as it cast loose its moorings from an industrial wharf in the upper reaches of the San Francisco Estuary, not far from the town of Hercules, and began to maneuver the twists and turns leading into San Francisco Bay proper, toward open ocean. Although it has since retired, the SeaRiver *Benicia* at the time was among the largest in a fleet of oil tankers operated by SeaRiver Maritime, Inc., formerly known as the Exxon Shipping Company. The ship takes its name from the town of Benicia, a suburb of strip malls and refinery towers set along a narrow channel between San Pablo Bay to the west and Suisun Bay to the east. The town of Benicia sits on the north side of the channel; the town of Martinez sits on the south side. A steel-girder highway bridge links them across the channel. The underside of that bridge was visible from two hundred feet below, atop the tanker's bridge. A crisp blue sky was visible too, as well as seagulls wheeling in mockery of the rush-hour traffic, which had ground to a standstill.

The SeaRiver *Benicia* was embarking on its regular run to Valdez, Alaska, to fill itself with crude oil from the great Trans-Alaska Pipeline. Fully laden, the vessel can carry slightly more than a million barrels—or about forty-two million gallons—of oil. The ship had arrived late the night before, having already off-loaded several hundred thousand barrels in order to raise its draft and easily sail the shallower channels of the upper Bay. In port, under floodlights and a brilliant half-moon, the crew, in company gear—hard hats, coveralls, gloves, safety glasses, steel-toed shoes—worked smartly about deck, attaching the elephantine pumps that would draw the liquid cargo from belowdeck tanks into a shoreside tank farm. All night the ship thrummed with the business of extraction. Around dawn a different set of pumps started up, drawing water from the harbor—which daylight revealed to be green and teeming with algal life—into a segregated set of belowdeck tanks for use as ballast for the return trip to Alaska. After several hours of pumping, the SeaRiver *Benicia* had gained sufficient ballast to set sail, and shortly after three o'clock she did so, bearing an assortment of organisms: twenty-one crew members; uncountable microscopic plants and animals—from diatoms and dinoflagellates to copepods and mollusk larvae—that had been drawn in inadvertently with the million-odd gallons of ballast water; two larval marine biologists and one incidental traveler.

The modern cargo ship is the world's preeminent vessel of desire. At any given moment some thirty-five thousand ships large and small are at sea, bearing supplies from one port to another: oil from Valdez to San Francisco, automobiles from Los Angeles to Tokyo, wood chips from Tasmania to Korea, Christmas trees from Oregon to Hawaii. Eighty percent of world trade is conducted by sea, and the ships have grown to impressive size. The largest container ships can carry six thousand boxcar-size cargo containers; port cities must dig deeper harbors and build larger facilities simply to handle them. They are monuments to consumption. The main deck of the SeaRiver *Benicia* was a hundred and seventy feet wide and nine hundred feet long, much of that space occupied by an orderly labyrinth of steel pipes. The ship's engine room, belowdecks, was a cavern of gigantic boilers, coolers, evaporators, and turbines. The propeller shaft alone was a yard wide. The din was such that the engine-room crew had to wear earplugs at all times. Only in the engine-control room, a glass skybox filled with computer terminals and instrument panels, was it possible to hold a normal conversation. Much of the time the engine room can run itself; the safety-automation program that guides it is common in nuclear power plants.

Running such a ship requires tremendous energy. An oil tanker the size of the SeaRiver *Benicia* consumes six hundred barrels of fuel each day; a large fleet of tankers may consume a hundred million dollars in fuel annually. The crew requires fuel too. Minimum requirements, spelled out in the Articles of Agreement that are signed by the captain and posted in the galley, mandate that a seaman "shall be served at least three meals a day that total 3,100 calories, including adequate water and adequate protein, vitamins, and minerals in accordance with the United States Recommended Daily Allowance." They receive this and much more. The abundance of food aboard the SeaRiver *Benicia* was staggering. At breakfast, through a small portal in the galley, a person could order eggs, bacon, sausage, pancakes, and/or French toast to one's heart-stopping desire. For lunch and dinner: pork chops, roast chicken, roast beef, beef stew, beef Stroganoff, spaghetti with massive meatballs, mashed potatoes, sweet potatoes, green beans, lima beans, rice, corn. Between meals, at all hours, there was the snack bar, a cornucopia of corn chips and pretzels, popcorn, doughnuts, fig bars, chocolate cake, peanut butter, melon slices, orange juice, lemon meringue pie. Of the challenges I would face in the coming days at sea, few were as daunting as the self-induced torpor brought on by the temptation of endless ingestion.

The rear quarter of the SeaRiver *Benicia* was devoted to the needs of the vessel itself. The engine and pump rooms were below; above, the house, a five-story superstructure, contained the crew: galley, crew's quarters, recreation lounge. The bridge, a large, open room with the latest high-tech navigation and communication equipment, as well as a panoramic view, sat at the very top. The vast remainder of the vessel, everything forward of the house, served the needs of the rest of the world. The tanker carried eleven cargo tanks in all, each one roughly the dimensions of a municipal swimming pool and sixty feet deep: three tanks on the starboard side, three tanks to port, and five tanks down the center, like a giant ice-cube tray. Four additional segregated tanks, two apiece on the port and starboard sides, held ballast exclusively. A layout of the tanks, pumps, and valves was visible on a lighted panel in a control room next to the galley; here, deck officers supervised the status of the cargo and ballast tanks and coordinated any change in their respective contents. At this moment in its voyage, the ship carried no oil. Instead, it carried ballast: nearly three hundred thousand barrels of Bay Area ballast water that, on arrival in Valdez, would be donated to Prince William Sound. Ballast water may be taken on, released, or shifted from tank to tank at any point during a journey—to tip the ship slightly aft to aid in the loading or unloading of oil; to raise or lower its draft and so optimize the use of its propeller and rudder; to keep the ship stable and maneuvering smoothly through changing seas. Ballast water does for the modern cargo ship what sandbags do for a hot-air balloon, but offering much finer control and demanding the employ of complicated mathematics. Once upon a time, a chief mate would refer to a manual filled with calculations—how much ballast to take on or release, depending on the weather, the ship's speed, and the amount of cargo and ballast already on board—and figure out the math in longhand. Now a sophisticated computer program handles the algorithms, allowing humans to devote closer attention to the numerous gauges, dials, and knobs in the ballast-water control room.

A decade ago, the average member of Congress had no more interest in ballast water than in hot-air balloon sandbags. By the mid-1990s, however, these officials had become fluent in the technical language of ballast. They had had explained to them the difference between a ship "in ballast" (carrying ballast water but no cargo) and a ship "with ballast" (containing ballast water and cargo), between the "ballast leg" and the "cargo leg" of a ship's journey. They had heard Jim Carlton testify several times on the subject of the damage caused by zebra mussels and other organisms introduced by ballast water. Ballast tanks, formerly viewed as mere aides to transportation, were now officially considered modes of transportation unto themselves—giant, moving aquariums that threaten to infest local waters with any number of exotic marine pests. Calls for ballast-water control became regular and urgent. The National Invasive Species Act of 1996 mandated that some form of nationwide ballast-water management plan be instituted, although there was no consensus as to which approaches or technologies to apply. The International Maritime Organization, with Carlton as a consultant, chimed in with suggestions and requirements. Experimental projects sprang up. Shipping companies like Arco, Mobil, and SeaRiver donated their ships, time, expertise, and money to examine and test the various possibilities. Clearly some sort of federal regulation would soon come into existence. The shipping industry faced a choice: either step aboard and help shape the policy, or sit still and watch the policy broadside them. Either they were on the boat or off the boat.

From the observational platform outside the bridge, I watched as our ship passed through San Pablo Bay and entered San Francisco Bay proper. Night had fallen and the City of Lights was visible off to port side, glittering and restless. A mile to starboard I made out a low silhouette of rock, Alcatraz, and recalled a quote from Samuel Johnson that the ship's chief engineer had posted on his cabin door: "Being on a ship is being in jail, with a chance of being drowned." The SeaRiver *Benicia* glided on toward the mouth of the Bay. The Golden Gate Bridge, astream with headlights and taillights, loomed ahead, then overhead, then dwindled from view behind us, leaving only the constellated sky: Cygnus, Draco, Hercules. Inside, the bridge was dark except for the glow of a radar screen and a table lamp illuminating navigational charts on the chart table. The captain himself, a tall man in a flannel shirt, stepped outdoors briefly and gazed back at the receding bridge. "Every once in a while I think about staying ashore," he said. "But then I listen to the traffic reports." We slid past darkened headlands, along a channel marked by green- and red-winking buoys. The aroma of eucalyptus slowly faded; the scent of sea-foam was ascendant.

A ship is a steel propagule, its ballast tanks filled with propagules of living carbon. Ballast-water regulation would undo this axiom. But how best to do so? Filter the water before it enters the ship? Heat it high enough to kill the organisms? Zap it with ultraviolet radiation? Require ships to exchange their ballast tanks at sea and purge their estuarine waters? Before those questions can be adequately answered, much more needs to be learned about the living propagules that undergo a ballast journey. Of those organisms that first enter a ballast tank, what fraction survives to disembark? What factors—temperature, salinity, length of voyage, type of organism—most affect their survival rate? Of those propagules that ultimately disembark, what fraction flourish and spread in their new habitat? In short, to what extent, in studied, quantitative terms, does ballast water and shipping contribute to the spread of marine organisms? And how—in studied, quantitative terms—does it do so?

Carlton offered a starting point with his Shipping Study, which had been mandated by the Non-indigenous Aquatic Nuisance Prevention and Control Act of 1990. Before long the law required more definitive answers from science; without them, any federal attempt to regulate or control ballast water would be pure guesswork. The chosen researcher this time was not Carlton, but a friend and colleague of his: Greg Ruiz, a marine biologist at the Smithsonian Environmental Research Center in Edgewater, Maryland, and a one-time Carlton postdoc. Carlton had been capital in steering science toward the notion that a ship is a moving island, a mobile home for marine organisms from around the world. Ruiz has joined the bridge now too, with an even more finely measured view of the horizon.

A ship and its ballast tanks are an ecosystem, an ecological environment as self-sustaining as a tropical atoll or a bookshelf aquarium. Any ecosystem can be approached with the same inquiry: What is the climate like there? How do conditions vary from day to day, from ship to ship, from tank to tank? Presented with a true geophysical island—one of the Mariana Islands, say, or a tract of Hawaiian rain forest—a biologist would begin to conduct field research on it: take measurements, plot characteristics, conduct experiments, detect patterns, make predictions, all by way of treading the elusive path from observation to hypothesis to general principle. He or she would camp out for days at a time and get to know the place. Carlton reached this archipelago first; Ruiz is arriving in force. He has established a lab—the National Marine Invasions Laboratory—to explore it, and for more than a decade has been sending young biologists to conduct fieldwork: to ride oil tankers and cargo ships from one port to another, across seas and oceans; to plumb the mysteries of the ballast tanks and their various international inhabitants.

So it was that I found myself aboard the SeaRiver *Benicia* in the company of two naupliar biologists from Edgewater. According to the protocol of the experiment, the ship would be traveling as much as two hundred miles offshore; the trip would last several days, weather depending. We would be temporary castaways on a high-seas island, assigned to comb insights from the ship's nonskid-surface shores, with only beef Stroganoff and beef stew and roast beef and chocolate cake and Fig Newtons to eat and a bottomless sea of coffee available to drink.

The bridge of the SeaRiver *Benicia* is a white steel box at the very top of the ship, well over a hundred feet above the ocean surface, with a sweeping view of the open sea. The bridge of the National Marine Invasions Laboratory, from which Greg Ruiz designs and steers a dozen or so ballast-water research projects, like the one under way aboard the SeaRiver *Benicia,* is a white metal box on a marshy inlet of Chesapeake Bay, with a view of the wooded campus of the Smithsonian Environmental Research Center. Most research projects at SERC are conducted within a complex of modern laboratories and office suites, or in one of the many carefully marked field sites in the surrounding woods or on the adjacent bay. Owing to a shortage of available space, however, and despite the fact that Ruiz's ballast research attracts more than a million dollars in annual grants from the U.S. Coast Guard and other federal agencies, much of his laboratory work is conducted in a white trailer home propped on cinder blocks near the edge of the woods. Fair to say that whatever ballast-water policies the federal government embraces in coming years—and the measures that the shipping industry ultimately adopts—will be guided largely by the science that emerges from this tin can. It struck me as an strangely apt setting for the study of global biological travelers: an immobile mobile home, a barnacle with aluminum siding. The first day I visited, a young lab technician sat inside at a workbench and, using tweezers, was struggling to pull apart an encrusted lump in front of him. He was dissecting barnacles. "They start out as cyprids; then they're barnacles," he mused. "One morning you look, and lo and behold, there's a tiny barnacle. It's like, Whoa, how'd that happen?"

Ruiz himself was stationed in a newer building nearby, in a small office he had moved into only days before. Among the lab techs at SERC there is an informal game that turns on a hypothetical question: In the unlikely event that Hollywood ever makes a movie about ballast-water ecology, which celebrity actors would be hired to play the lead characters? So far, consensus has it that the role of Jim Carlton would be played by Richard Dreyfus: a compact, energetic effusion of humor, goodwill, and scholarship, not unlike the marine scientist in *Jaws*. The role of Greg Ruiz, all agree, would go to the actor Jeff Goldblum: tall, dark-haired, wryly enigmatic, his loping manner disguising a pacing and preoccupied intellect—think *Invasion of the Body Snatchers*, *Jurassic Park*, or *The Fly*. At the moment, Ruiz was reclining in his swivel chair with his feet on his desk, exhibiting a battered pair of running shoes. The records of resettlement—cardboard boxes, manila file folders—surrounded him in heaps. The phone buzzed; he ignored it.

Formally trained as a zoologist at the University of California at Berkeley, Ruiz was one of three postdoctoral researchers who worked under Carlton at the Oregon Institute of Marine Biology. "He's one of the people who arrived as one thing and left as an invasion ecologist," Carlton says. Some of Ruiz's colleagues might question the depth of the zoologist's metamorphosis—"He stumbled into this invasion thing, he's just in it for the critters," says one—but few would dispute his contribution to the burgeoning field of ballast-water ecology or his central role in the evolving federal policies concerning ballast-water management. "Because this work involves ships and commerce, it has local, regional, and international ramifications," Ruiz said at his desk. "Everything we're doing is of interest from a management perspective. I've been interacting more with agencies like Fish and Wildlife that are implementing policy or advising on how to do it. I've become more aware of how those decisions are played out. But invasions also provide an excellent opportunity to study some basic tenets in marine ecology, like larval dispersal and community structure. That's one of the big issues in ecology: What is larval supply? Is a particular ecosystem a closed system, where the adults supply the larvae for the next generation, or is it an open system, where there are new larvae coming in and mixing? Over what distance are the larvae dispersing?"

As the director of a research lab, Ruiz manages a staff of two dozen and designs and oversees the research protocol: who will investigate what and when, how they'll do it, and how much it will cost. There are grant proposals to write, congressional subcommittees to address, potential funding agencies to impress and cajole. There are professional conferences to attend, talks to give, meetings to arrange with various federal officials. In addition, he is the academic adviser to a dozen or more undergraduates or doctoral candidates, all of whom have spent some stage of their academic life cycle at SERC working on ballast-water issues. Fifteen years younger than Carlton, Ruiz is already supplying larval biologists for the next generation and tracking their dispersal.

That is only the logistics; then there is the actual science. Along with the ballast-water experiments and studies, Ruiz pursues his own fieldwork: for several years now he has studied the incursion of the European green crab into several non-European environments, including estuaries in northern California and Australia. His research results are notable for their scientific depth and their breadth of view, and they have been published in all of the prestigious science journals. Presumably, he writes these articles while traveling, since an airline seat is one of the few places Ruiz can be found sitting for longer than thirty minutes at a stretch. Mile for mile, sleepless night for sleepless night, Ruiz is an even match for Carlton.

At heart, Ruiz aspires toward a sort of grand numerical synthesis. The occurrence of invasions in marine ecosystems has been acknowledged and studied for years now, as has the role of ships and ballast water in the transfer of introduced marine organisms—by Carlton, in both cases, but by many other researchers as well. What is not well established, at least not to Ruiz's satisfaction, is precisely how those two phenomena relate to one another. How many organisms are being borne in ballast water, and to where? What fraction of those organisms successfully invade? How and to what extent do those invaders affect their new environments? "He is connecting the dots between rates and impacts," Carlton says. Although the numbers of biological invasions sound ominous, in fact marine scientists have only begun to explore their actual implications. Ruiz, in his office, said, "We don't know what the impacts of most invaders are.

There are a lot of inferences, but not much data. If you want to ask how frequently things invade San Francisco Bay, in terms of rate, we don't know. You can look at Jim's list of two hundred and twelve species and say, 'That's a lot.' Are they having an impact? Probably. But there aren't enough data. And most of the data are correlative. It's not sufficient to prove cause and effect. Especially in estuaries, where there are a lot of anthropogenic effects that often occur simultaneously with invasions: nutrient loading, pollution, sedimentation."

In the pursuit of quantitative information, Ruiz has developed the National Ballast Information Clearinghouse, a comprehensive daily record of the floating ecosystems that enter every major port across the country. With help from the U.S. Coast Guard, critical information about all ships arriving in their respective ports is fed into a computer database in the immobile home of Ruiz's laboratory: their size, their cargo, their last port of call, the amount of ballast water they carry. The accumulating database offers a more accurate way to probe the link between propagule pressure and invasion rate. For example, although Chesapeake Bay is one of the nation's largest recipients of ballast water, taking on some twelve million metric tons annually, it has fewer documented introduced species than San Francisco Bay. Several years of National Ballast Information Clearinghouse data might help clarify whether the Chesapeake's lower invasion rate is a product of real differences in shipping behavior or the environment, or whether it merely highlights the added scrutiny that San Francisco Bay has received over the years from Jim Carlton.

On occasion, as part of a separate research project designed to get a fix on the kinds of ballast organisms that come to Chesapeake Bay and the conditions under which they arrive, the lab techs make house calls. They drive to the docks in Baltimore and Norfolk, board ships in port, collect samples of whatever lives in the ballast tanks, and record the physical attributes of the habitat: the temperature, salinity, and dissolved-oxygen content of the water at various depths within the ballast tanks, as well as the temperature of the Chesapeake water into which the incoming ballast organisms would likely be released.

The previous day, Betsy von Holle and Scott Godwin, two of the lab techs, had called on the *Edridge*, a Dutch bulker that had docked in Baltimore to load up on coal. The ship was considerably smaller than the SeaRiver *Benicia*, with nine cargo holds capped by sliding steel lids. Although the holds had arrived carrying ballast water, all but one had already been emptied into the bay. Von Holle and Godwin were directed to a round, open hatch on deck that offered access to it.

Typically, very large organisms are prevented from entering a ship's ballast tank by a grate that covers the pipe that draws in water from outside. Nonetheless, SERC researchers have found things living in ballast tanks that were considerably larger than microscopic: crabs, eels, even a school of mullets. Aboard one ship, crew members boasted of sitting around the edge of an open ballast tank with their fishing poles and catching lunch. Aboard the *Edridge*, I could make out the surface of the ballast water, several feet below the open hatch, illuminated by a shaft of sunlight. The water was green with the chlorophyll of phytoplankton and had a grainy appearance. It took me a moment to notice that the grains were moving and were doing so under their own power. They were tiny, swarming copepods, Linnean cousins to crabs, lobsters, shrimps, and the other hard-shelled members of the class Crustacea. A few thousand copepods might fill a thimble. The number of thimblefuls in the *Edridge*'s ballast tank was beyond calculation. For a moment I had the dizzying sensation of standing on the brink of two teeming worlds: the titanic realm of ships and berths and cargo cranes, the world of global commerce against which my own body seemed dwarfed in size; and the Lilliputian realm of plankton, brimming and busy, of which I could see but a tiny fraction through the open hatch of the tank.

Dipping a plankton net repeatedly into the tank, Godwin and von Holle managed to fill several clear plastic bottles with *Edridge* ballast water and its inhabitants. They took a series of measurements, then brought the bottles back to the SERC trailer and set them on a lab bench. Then they and two other colleagues sat down at microscopes stationed around the room and went at it, peering at small samples of the water, trying to identify the copepods—and whatever else might be in there—taxonomically. At one point Ruiz stopped in to check on progress. He noticed the bottles, which were astir with grainy copepods. "Wow! That's gotta be the densest concentration I've ever seen." He picked up a bottle for a closer inspection of its motes. "That one looks like it's got egg sacs. That one looks like the larvae of a decapod, maybe a crab. That one looks like a sea monkey. It looks like a bunch of sea monkeys!" He paused to considered that thought. "So all these suckers are out in the bay now. Some of them are probably doing pretty well."

Though small in size, copepods are highly instructive organisms in ballast-water research. They are the most abundant form of zooplankton in the world's waters, both fresh and marine. As a consequence, they are the most abundant form of life in the water of any given ballast tank in the world. Their name in Greek means "oar-foot"; they are minuscule argonauts, propelling themselves by scrabbling through the water with their tiny legs. To quickly gauge the "productivity" of any given ballast tank—the ability of its waters to support life—one need look no further than the copepods rowing in it. Are they many or few? Are they spry or sluggish? The more productive the ballast waters, the more likely its inhabitants will survive from one foreign coast to another, copepods and non-copepods alike. In addition, there are countless individual species of copepods—the many species native to the Amsterdam harbor are distinct from the copepods of the Chesapeake, which are different again from the copepods in San Francisco Bay, which are different again from the copepods of the salty high seas. In theory, the copepods in a ballast tank could serve as a sort of fingerprint of its waters and provide a ballast-water inspector a quick determination of whether a ship had properly exchanged its ballast in mid-ocean according to regulations.

The trick is figuring out which kind of copepod you are actually examining. These microscopic animals are divisible into at least ten suborders, each one with its own vast subset of copepod families, genera, and species. Are you looking at a monstrilloid copepod, of which there are approximately eighty species, or a mormonilloid copepod? Is that a platycopioid or a poecilostomatoid? Paul Fofonoff, a research assistant in Ruiz's invasion lab, took a bottle of *Edridge* ballast water and poured a small amount into a shallow glass dish. He slipped the dish under the oculus of a microscope nearby, pulled up a low metal stool, sat, and peered down through binocular eyepieces. "Let's see what we have here!"

Fofonoff is something of a curiosity at SERC. His main project involves compiling a history of biological introductions in the Chesapeake Bay, based on historical records, museum collections, and a thorough search of the scientific literature. He possesses what may well be a photographic memory. As one colleague put it, "His penchant for information is excrutiatingly detailed. We've learned to ask him for the short answer." When I asked him one afternoon for a brief overview of his findings, he replied readily. "There are thirty species of submerged aquatic vegetation in Chesapeake Bay, of which six are introduced," he began, and continued—virtually without pause—for an hour. "That includes aquarium plants. The first invasion was curly pondweed; it was first seen in the Potomac in 1876. The first one to cause trouble was the water chestnut. It came from Asia as an ornamental garden plant. It was first seen in the Concord River in the 1800s; then it spread to Lake Ontario. It was seen in the Potomac in 1923. It's a floating plant; it causes problems with navigation. It was pretty well eradicated by the 1930s. In 1983 . . ." When at last I rose to leave, Fofonoff handed me a stack of photocopied scientific papers, a small fraction of the voluminous research he has absorbed. I read them and, two days later, went to his office to return them. His recitation started up precisely where it had ended, as though I had never left. "In fishes, hybridization is common in disturbed communities . . ."

It seemed possible that he could expound, in detail, on anything: Asian rice weed, the narrow-leafed cattail, European minnows in Philadelphia. Now, with a dish of ballast water from the *Edridge* in view in his microscope, he added, "A lot of the organisms that we know are being transported in ballast water, like copepods, often are little known in terms of taxonomy and have a poor historical record." He looked up from the microscope, roused from the aqueous underworld. "It wouldn't be surprising if several copepod species were in there. But it will take a specialist to be certain. If somebody told me this ballast was from Boston Harbor, I wouldn't know any better just by looking."

Next to us at the workbench, von Holle was peering into a microscope, struggling to make sense of the magnified hash of copepods. Textbooks and identification manuals were propped open around her on the workbench: page after illustrated page of feathery legs, sticklike antennules, multi-segmented thoraxes, and other allegedly distinguishing copepod characteristics. Uppercase and lowercase letters marked the most important anatomical features in the illustrations. On other pages, the uppercase and lowercase letters were rearranged into elaborate charts: by cross-referencing the various letters, a reader could inch her way toward the species name of the copepod she presumably was viewing under the microscope. Von Holle proceeded in this manner, examining the charts and illustrations, peering into the microscope, then flipping again through the pages. With one hand she drew sketches of her quarry that rivaled in precision and elegance the illustrations in the books. These would be filed away in a binder as an additional visual reference for the other lab techs.

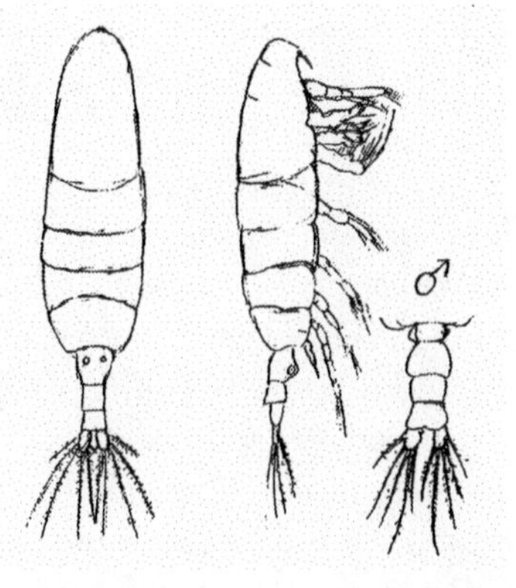

I was impatient to look. Fofonoff scooted aside in his stool and directed me to his microscope. Evidently, copepods possess only a single eye on what passes for a forehead. I confess that I was unable to make out any such facial feature. Fofonoff said that for the most part, the lab techs need not identify the copepods all the way down to the species level; it was enough to differentiate between one of three orders: calanoids, harpacticoids, and cyclopoids. I could not make out any physical characteristic that might distinguish those three categories. Nothing I saw looked like anything von Holle was drawing in her sketchbook. Instead, what I saw looked like a heap of transparent fleas: bare outlines of multilegged but otherwise shapeless creatures that hopped in and out of view too quickly to scrutinize. I drew a sketch in my notebook to later remind me what I'd seen.

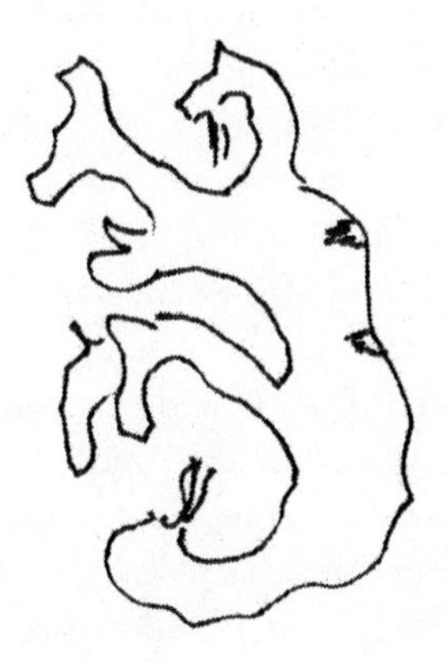

When Fofonoff informed me that the order Cyclopoida is further divisible into numerous genera including *Cyclops, Microcyclops, Macrocyclops, Megacyclops,* even a *Metacyclops,* I began to wonder if there weren't also some tiny seamen down there trying to escape, disguised in the fleece of sheep. "I think there are calanoids here; it's hard to see," he said. He had resumed command of the microscope. "I do see an adult female. I see a harpacticoid. I see a barnacle cyprid. I see a diatom . . ."

Across the campus in his disordered office, Ruiz said, "Those copepods could potentially become established here. It strikes me that they have a good opportunity to invade. The salinity is a good match. The temperature is pretty good. If it doesn't become established, then why not? You can get a lot of inoculations that don't result in invasion. Our lab studies suggest that it's not because the organisms aren't in good condition. So maybe it's the conditions in the environment: water quality, or the shock of the water temperature, or maybe the food isn't appropriate."

Another reason for Ruiz's interest in copepods is that, like ships with rats or like ship rats themselves, these tiny, oar-footed organisms can ferry troubling pests of their own, among them *Vibrio cholerae,* the bacterium responsible for cholera. Of the one hundred thirty-nine known strains of the *V. cholerae* bacterium, only two—types *01* and *0139*—actually cause the cholera disease. (Several other strains of *V. cholerae,* as well as other species of *Vibrio,* can cause milder diarrheal diseases in humans.) Increasingly, many scientists believe that cholera bacteria can be and have been spread by ballast water. In July 1991, six months after a cholera epidemic struck Peru and several neighboring nations—the first-ever cholera epidemic in South America—an identical strain of *V. cholerae 01* turned up in oyster beds in Mobile Bay, Alabama. Researchers sampled the ballast, bilge, and sewage water of nineteen cargo ships docked at Alabama and Mississippi ports; they found the toxigenic *01* strain on five ships, all of which had last stopped at a port in South America or Puerto Rico. Some of the ballast water had not been exchanged in months; evidently the bacterium can survive on ships for extended periods, in a wide range of temperatures and salinities. In the past century, cholera has rapidly spread several times from India, where the bacterium was first identified, to other parts of Asia, Europe, and the Americas. Scientists now suspect that ship traffic may have been at least partly responsible for the spread.

Recently, Ruiz, in collaboration with microbiologist Rita Colwell, had begun inspecting the ballast water of ships arriving in the Chesapeake for evidence of *Vibrio* bacteria. How often does the toxigenic strain appear in ballast water taken from ports where a cholera outbreak has recently occurred? In what density do the bacteria occur? Simply finding the bacteria is a challenging task. They are not visible under standard microscopes, and they often exist in a form that cannot be grown in laboratory petri dishes. However, high concentrations of *Vibrio* bacteria often can be found attached to other aquatic organisms: duckweed, green algae, and—most notably to a ballast-water biologist—copepods. It may be that *Vibrio* bacteria gain nutrients by dining on the chitinous bodies of copepods. Perhaps, thus bound to the animal, the bacterium also stands a better chance of surviving the acidic hazards of the human stomach and is thus more likely to cause disease. A better understanding of the copepod-cholera connection might offer insight into ways to check the spread of the disease.

"If you're going to talk about some sort of treatment or management of ballast water, bacteria and viruses should be included in the equation," Ruiz said. "It's not adequate to talk about what will kill a dinoflagellate or a crab larvae or a copepod. What is an effective treatment for microorganisms?" In the meantime, he said, it was Fleet Week in Baltimore harbor for another shipload of ten-legged foreign sailors. "That inoculation"—he jabbed a thumb in the direction of the trailer-home lab and its crowded jars of copepods—"is spectacular. If you were a copepod, you couldn't ask for a nicer opportunity to move to Chesapeake Bay with hundreds of millions of your closest friends."

20

Two p.m., November 8. The SeaRiver *Benicia* is four days out of San Francisco, two hundred miles off the coast of Oregon. On this day there is nothing to see but the sea: hazy, slate gray, a field of whitecaps stretching to the horizon. At this time of year, a menacing low-pressure system nicknamed the Siberian Express regularly races across the northern Pacific from Kamchatka to North America. Today, however, there is only a slight, steady northwest swell, ten to fifteen feet, that keeps the ship on a sleepy roll. Viewed from the wheelhouse, against the vast acreage of the vessel, the waves appear small, benevolent—until a crew member, a minim in rain gear and hard hat, appears on deck and renders humanity in vulnerable contrast.

The ship's safety regulations are simple: during the day, no one goes on deck without notifying the bridge; after dark, no one goes on deck, period. If the daytime sea grows too rough—if green seawater begins to wash over the bow—the deck is closed to all hands. The reasoning requires no explanation among the crew. If a person fell overboard, it would take the ship seventeen minutes to stop, reverse course, and return to the point of the accident—provided someone saw him go over. The temperature of the water at this latitude approaches fifty degrees Fahrenheit. Unprotected, a person like me could survive for thirty minutes before succumbing to hypothermia: seventeen minutes to wait for the ship to maybe return, thirteen more for the bridge, miraculously, to spot my melon-size head on a seascape of towering whitecaps and send out a lifeboat.

I was essentially alone with my morbid preoccupations. Safety drills and training are such an inherent part of a crew's work life that the sea exists for them as a manageable hazard, neither ignored nor distracting, like the sight of blood to a surgeon. "Second nature," as one put it. I was still grappling with first nature. One morning in the galley I saw a notice on the calendar announcing a mandatory and regularly scheduled drill: "Survival Suit Try-on Day!" To extend survival time in the event of a capsized ship, every crew member is outfitted with a specially designed survival suit. After breakfast I returned to my cabin and extracted my suit from a duffel bag in the closet. It resembled a full-body wet suit, made of thick insulating neoprene, and was a blazing orange color. Its instructions said I should slip into the suit as if stepping into a pair of coveralls, but I soon found myself wrestling my way into it on the floor, a project I doubted I would have the patience or wherewithal to undertake in the dire circumstances that might require it. Finally I got it on. I was wholly cocooned, my unshod feet in booties, my hands in three-pronged mitts; in the mirror I looked like a bright orange Gumby. I zipped up the suit and pulled a snug hood over my head, and was instantly overwhelmed by claustrophobia. My heart rate doubled; I understood then the terror of the straitjacketed. It was unclear how long the suit would keep me warm and afloat: a few hours, perhaps a day. I tried to imagine—indeed, in subsequent days I could not stop imagining—bobbing on the sea in it, waving like a red copepod at a dwindling ship. I pictured the hours alone amid towering waves and an undulating, unbroken horizon. Hypothermia sounded like a relief.

Such thoughts did not appear to preoccupy Kate Murphy, one of the two SERC biologists on board, as she made her way across the ship's rolling deck. She crouched near a round hatch that opened onto the forward ballast tank on the ship's port side, a tank formally designated as Two Port. Her deck chores as a ballast-water researcher kept her on a triangular path between the hatches of three ballast tanks: Two Port and Four Port, both on the left side of the ship; and Four Starboard, on the opposite side, across a walkway that ran the length of the deck and was known as Broadway. Although the sun was bright, Murphy wore heavy-weather rain gear to ward off the wind and the luffs of spray that curled over the bow with stinging regularity. A large plastic wheelbarrow was parked beside her, firmly lashed with rope to a deck pipe. It was piled with laboratory equipment: bottles and jugs of various sizes and colors for collecting samples of ballast water; a length of rope; a large plankton net resembling a wind sock; scissors, pliers, flashlight. Murphy was busy filling specimen bottles with ballast water, which she drew from the tank by means of a slender plastic tube attached to a suitcase-size pneumatic pump. It was trickling work, prone to spillage, and Murphy's hands had turned red from the wind and cold.

The voyage of the SeaRiver *Benicia* would mark one in a grand series of experiments designed by Greg Ruiz, each one carried out by SERC lab techs aboard a different supertanker. Their aim was not to examine the array and condition of organisms riding in ballast tanks, which by this point had been exhaustively documented by others. Rather, these experiments would test the proposed method of ridding the tanks of their inhabitants. With the Great Lakes for a model, federal regulatory agencies have begun to rally around mid-ocean exchange as the stop-gap solution: before entering an American port, all ships would be required to empty their ballast tanks while still at sea and refill them with high-salinity ocean water. Presumably, this would kill or flush out any estuarine organisms picked up at the port of departure and replace them with oceanic organisms, which would be unlikely to survive release in the next estuarine port. One virtue of this approach, in addition to its alleged efficacy, is its low cost: no retrofitting of ships with expensive ultrafine filters or ultraviolet radiation machines; no chemicals of questionable environmental safety would be required. A ship merely has to do what it was already designed to do—empty and refill its ballast tanks. It would simply do so at sea rather than in port.

Imagine now that mid-ocean exchange has become required by law nationwide (currently it is required in some ports, and is a strongly recommended voluntary exercise in others) and that an arriving ship claims to have obeyed it. How would a coast guard inspector know for certain? The SeaRiver *Benicia* experiment in effect would test several methods of fingerprinting the water in the ballast tanks. Over the course of several days, the chief mate would empty and refill three tanks—Two Port, Four Port, Four Starboard, the three points on Kate Murphy's triangle—as many as three times. Murphy and her fellow traveler, Brian Steves, would collect water samples from each tank before and after each exchange and at the beginning and end of the trip. Afterward, all the samples would be sent back to Ruiz and scrutinized for the presence, in varying amounts, of various potential tracers, some of which had been added to the ship's tanks in San Francisco: rhodamine dye, dissolved organic material, trace metals, certain species of phytoplankton, different isotopes of radium. Hopefully, one of them would prove useful in showing beyond doubt that the SeaRiver *Benicia*'s ballast water had been thoroughly exchanged, or nearly so, at sea.

For Murphy and Steves, this would add up to several days of nonstop scurrying on deck, hatch to hatch, tank to tank, with a wheelbarrow of equipment in tow. My assigned task was to write down equipment readings as they were called out. As experimental protocols go, it was classic Ruiz: statistically exacting, designed to gather as much information in as short a period with as many controls and built-in redundancies as possible, and physically exhausting. For a staging area, Murphy and Steves had taken over a small stateroom. The room was soon awash in field gear and scientific equipment: boots, rain gear, helmets, safety glasses, boxes, crates, bins, jars and bottles, labels, tapes, straps, rope, tubing, wrenches, notebooks and data sheets, waterproof pencils—on the bunk beds, on the floor, hanging from hooks, heaped under the sink.

Venturing on deck to collect one set of samples was a three-hour expedition. Begin at eight a.m. with the coveralls, rain gear, hard hat, and other safety gear required for crew members. Load up the wheelbarrow with the crates of jars and bottles and the suitcase-size pneumatic pump. Totter out in the wind toward the bow, to Two Port, Hatch A. Attach the pump to an on-deck air hose (for power), then to the slender plastic tube marked "1 m" poking its head out of Hatch A, and begin collecting ballast water from one meter down. Fill a small amber medicine bottle, close it with a black cap; that sample will later be analyzed in the lab for the presence of rhodamine dye. Fill an amber medicine bottle, close it with a green cap; that sample will be analyzed for dissolved organic matter. The trace-metals sample goes in a small plastic bottle. Remove the tube marked "1 m" from the pump, attach the tube marked "10 m," and collect the same ballast samples again, this time from ten meters down in the tank. Drop the conductivity meter—a large, plastic, expensive piece of equipment that records salinity, temperature, and dissolved oxygen—on a rope to a depth of one meter down in the tank, haul it back up, jot down the readings, repeat at twelve and twenty meters. Pack up the equipment, load up the wheelbarrow, totter over to Hatch B, unpack, repeat. At eleven-thirty, quit for lunch. Entering the galley, I felt as though I had spent the day skiing: red-faced and weary, drained from the mere effort of holding myself upright in a steady wind.

It was possible during those long hours on deck to forget one's place in the larger world—which, given the lack of recognizable landmarks or any marks of land whatsoever, was essentially nowhere. There was the ship, and there was the sea rolling seamlessly, ceaselessly past it. The edge of the deck was marked all the way around by a low steel guardrail that served as a barrier between civilization and the marine wilderness. I quickly came to regard the railing as a kind of existential horizon. On this side, on deck, was the tangible world, the realm of urgent tasks and palpable objects, of mechanics and progress and knowns. On the far side, the unknowable, its implications too chilling and abstract to ponder for long. But as I said, that was just me.

I willed myself to focus my attention downward—on the ship's deck, painted red with a special nonskid coating that reminded me of my high school running track; on the pneumatic pump and the accumulating collection of water-filled bottles; on the round ballast-tank hatch that opened like a manhole onto the entrained sea below. The surface of the ballast water lay in darkness several feet below, out of sight, but it was audible, sloshing and smacking against the steel shores of the tank. In addition, the steady roll of the ship created an alternating pressure differential in the tank, such that the open hatch behaved like a blowhole: bellowing foul spray one moment, gasping inward the next, howling and wailing like Scylla and Charybdis. To collect the ballast-water samples safely, Murphy had to time her actions carefully with the respirations of the tank. At one point, a sudden upgust caught the plankton net just as Murphy was hauling it out of the hatch. The wind blew the net into the air like a wind sock and sent a plastic bottle skittering down the deck, then pinned the bottle against the guardrail. Murphy pursued it gingerly, trapped it with her foot, picked it up, and tottered back against the wind. She resumed her station by the ballast hatch, which had never ceased its rhythmic shrieking. "It's like the depths of hell down there," she said. "I'd rather fall into the sea than fall in there."

Austin James, the chief mate and the crew member whose principal responsibilities included emptying, refilling, and generally managing the SeaRiver *Benicia*'s deck and pumping operations, kept a watchful eye on the research activity from the towering height of the wheelhouse. "I'll be very interested to see what happens, how the experiment turns out," he said. For James, managing the ballast for a ship this size is an exercise in minimizing stress—the ship's and, by extension, his own. He directed my attention out the window, far up the deck to the forecastle. The bow was moving: rising and falling with the waves as well as flexing on its own ever so slightly. The ship's hull, he said, was carefully constructed to torque and flex to absorb the full and often conflicting forces of the sea. Some flexing is good, but not too much. Even in fair weather a tanker should ride deeply enough and slowly enough through the waves that its bow never hangs unsupported in the air. Some ships are specifically built for grace, with narrow bows that carve through waves more smoothly and with less effort. Not this one; it sported a rectangular design to facilitate the movement of oil in bulk. "We're a block," James said. "We're a big block going through the water."

The stress exerted on a ship is calculated according to a complex algorithm that considers the ship's size and tonnage, the amount of ballast and cargo on board, and such variables as weather, wave height, and the ship's speed. In a nutshell, the larger the ship and the larger the sea, the greater the potential stress exerted on the hull. The full scale of this stress had been impressed on me some months earlier at a scientific conference during a slide show presented by a representative of the U.S. Coast Guard. The slides were photographs of the forward deck of a large cargo ship in violent seas, as seen from the bridge. Or not seen: the forward deck was barely visible, owing to the enormous waves, fifty to sixty feet in height, that broke across the bow and buried the deck under a wash of green and white. The speaker had aimed his presentation at the biologists in the audience, to help them appreciate the challenges of mid-ocean ballast exchange. Other than adjusting speed and course, adjusting ballast is the sole means of minimizing the stress on a ship as it moves on the open ocean. At all moments during the SeaRiver *Benicia*'s voyage, James would stay watchful of the sea and the weather forecast and would recommend accordingly slight adjustments in speed, course, or the uptake or release of minor amounts of ballast water. Suffice it to say that altering one's ballast water in the open ocean, whether for regulatory or experimental purposes, even on a sea as flat as a millpond, is not a task undertaken idly.

"I have to watch the stresses as we do it," James said in the wheelhouse. For the sake of Ruiz's experiment, the ship would try out two different methods of ballast exchange. In one, called an empty-refill exchange, a tank is first emptied of its ballast and then completely refilled. The second variety, called a flow-through exchange, draws in ocean water from below while the tank is still full and lets the old water burble out through the hatch onto the deck and back into the sea. Both methods take several hours, and neither is entirely free of cost. To accommodate all the ballast exchanges in Ruiz's protocol while minimizing the stresses to the SeaRiver *Benicia* would add a full day to her journey. With the ship burning roughly six hundred barrels of oil a day, at the then-price of thirty-one dollars per barrel, the cost of ballast exchange for one tanker on one voyage would add up to nearly nineteen thousand dollars. That is not an insignificant sum, even to a large company. An operation with thirty ships in its fleet (SeaRiver has fewer than a dozen), each making thirty runs a year, could spend more than fifteen million dollars annually conducting ballast exchanges—not including such hidden costs as extra crew time and the added wear and tear on a vessel.

For a chief mate, however, the principal hesitation to the notion of mid-ocean ballast exchange is its potential effect on the ship. At sea, the abiding law is the captain's law, and if for whatever reason—rough conditions, an overburdened ship—the captain determines that ballast exchange poses an undue risk to his cargo and crew, that is the end of the discussion. According to Ruiz's schedule of experiments, the SeaRiver *Benicia* had been due to empty and refill Four Port that morning, but the plan was abandoned when James determined that under the current sketchy weather, emptying the tank would expose the ship to 108 percent of the allowable stress risk. "The problem with ballast exchange—well, we've already run into it," he said. The flow-through method poses less risk—since the tank is never empty of ballast—and was permissible within the ship's stress limits. James watched from the wheelhouse now as water bubbled out of the Butterworth hatch of Two Port, spilled across deck, and rejoined the sea. He put the best face on it that he could.

"There's not nearly as much stress in a flow-through exchange," he said. "The liquid level stays the same. Your trim and stability isn't going to change." In the future, he added, ships could be specifically designed so that a flow-through exchange wouldn't interfere with deck activity. The flow-through method offered one additional advantage not to be overlooked, James said. Its stress on the ship was sufficiently minimized that the process could be semi-automated and would thereby occupy less of a chief mate's precious time.

21

By all accounts, Ruiz runs a successful ship in Edgewater, Maryland. Perhaps too successful: each new federal grant anchors the marine biologist more firmly to dry land. From within an aluminum-sided exoskeleton, he spends his days navigating the channels of government financing and grant requirements. His office does not have a window; if it did, it would provide a panoramic view of an endless sea of paperwork. One might say that as the movement and management of ballast water places stress on a ship and its operator, so the management of a ballast-water research laboratory places stress on the lab captain—the effects of which are not lost on a young crew.

"The great thing about Greg is, he's always thinking on a grand scale."

"He's fascinating. He's where we'd all like to be."

"He brings in the money. If he puts his name on a paper, it gets funded."

"It must put a lot of pressure on, feeling like you have to come up with all the ideas."

"He needs a manager."

"I feel sorry for the guy. He's a real field biologist; he's not a desk jockey. Yet he spends ninety percent of his time doing administrative work."

"He's a good example of what not to become."

Late at night on the bridge of the SeaRiver *Benicia*, Kate Murphy said, "He's a good businessman. Which means he's doing something he's not that interested in. Sometimes I wish he'd just cut out everything but three people and do what he wants." Something like this thought has occurred to Ruiz as well. From his windowless bridge in Edgewater he said, "I'd like to get out in the field a lot more. But then the number of people you take on in the lab—the scope of the program—can't be as big. That's part of the allure of California: it's an exceptional opportunity to look at invasion—and for me to do a lot of fieldwork."

California, for Ruiz, means Bodega Harbor, a small coastal inlet alongside Bodega Bay, a former cannery town an hour's drive north of San Francisco. One day, taking a break from Jim Carlton and Andy Cohen's biological expedition around San Francisco Bay, I drove up the coast to visit Ruiz in his native element. At seven o'clock on a cloudless morning it was easy to understand the attraction. The tide was low and dropping, exposing a sweeping plain of mud that gleamed in the early sun. The fragrance of brine and algae and marine efflorescence melded with the cool remnants of a lifting fog. Small fishing boats from the Bodega Bay wharves chugged across the tidal plain on a slender ribbon of blue water that passed through a rocky breakwater several hundred yards out and then to sea. Ruiz was walking briskly across the mudflat in knee-high rubber boots; dressed in jeans and a T-shirt and carrying a large white pail in one hand, he had the look of a man on his way to a day at the beach. I struggled behind in a pair of borrowed boots, sinking several inches into the muck with one step, then struggling with the next to extract my foot from the slurping gravity of tidal mud. Here and there, wide mats of a crepelike seaweed provided a more reliable walking surface. The acres of open mud bore countless small mounds generated by ghost shrimps tunneling through the mud just below the surface. Their subterranean network was so pervasive that it rendered the top several inches of mud essentially porous. Here, the distinction between land and sea became obsolete; it was a kind of nether terrain in which a hesitant pedestrian could quickly become mired.

"There's a pretty extensive underground system," Ruiz said. "If you walk out, you'll start to get sucked down. So don't stop."

As interested as Ruiz is in the organisms that are carried aboard ballast tanks and the physical conditions that prevail there, it is "the downstream end of invasion" that interests him most intensely: the subsequent fate of those organisms once they are released back into the environment, their efforts to establish themselves in their new territory, and the wider impact of their doing so. The recurring central character in Ruiz's fieldwork is *Carcinus maenas*, the European green crab. Formally native to Europe, the green crab has dispersed with human traffic to several corners of the world. It was first noted on the U.S. East Coast in the early nineteenth century, long enough ago that it has become "naturalized"—that is, most people have forgotten that in fact it is an introduced species. Shellfishermen view it as an underwater locust. The green crab single-leggedly crushed the soft-shell clam industry north of Cape Cod, reducing the annual harvest from 14.7 million pounds of clams in 1938 to 2.3 million pounds in 1959. In 1989 the green crab appeared in San Francisco Bay. In a kingdom of nonindigenous species, it quickly achieved majesty, feasting on native crabs and introduced Manila clams with equal fervor. More so even than the zebra mussel, the European green crab represents the kind of biological hazard that ballast water can easily transport around the world. Some marine biologists have taken to calling it the brown tree snake of the sea. In 1996 the green crab appeared for the first time in the coastal waters of Tasmania, Australia. And it has moved gradually northward from San Francisco, in 1998 and 1999 appearing in coastal estuaries of Oregon, Washington, and British Columbia. Ruiz has not yet detected evidence of the green crab in Valdez, Alaska, but it was within reason to imagine that at that very moment there was a ship under way, its ballast tanks brimming with San Francisco Bay waters, bearing crabby European immigrants above the forty-eighth parallel.

I must admit I was slow to appreciate the ecological danger posed by the European green crab. A snake I could grasp, literally and figuratively: sleek, cryptic, with a mouth capable of engulfing prey several times larger and a significant mythos attached to its tail. By comparison, a crab looks ridiculous. With ten twiggy legs, an inflated shell of a body, and two pinheads on stalks for eyes, it possesses all the grace of a windup toy and all the mystique of a cockroach. But that is appearance. Like the brown tree snake, the European green crab is an unusually irritable and rapacious creature. It owns a large and powerful pair of forward pincers, which it wields like can openers to pry apart oysters, clams, native shore crabs, and other hard-shelled delicacies. One study found that the common East Coast periwinkle *Littorina littorea* crawls for cover at four times its normal pace when it smells a green crab in nearby waters; within ten minutes nearly all the periwinkles in the vicinity will have escaped from sight. This ten-minute escape window is likely an evolutionary adaptation, as a green crab can eat a periwinkle in nine minutes and fifty-four seconds on average.

The green crab can thrive in a wide range of temperatures and salinities; it grows quickly and reproduces in quantity, producing up to two hundred thousand eggs at a time. On one occasion Ruiz showed me a small, bubbling aquarium he kept in the laboratory to hold adult green crabs. Perhaps a dozen were in there, scrambling on top of one another, nudging each other out of the way, drifting up on the artificial current like hard-shelled zeppelins. The native grapsid crabs of Bodega Harbor grow no more than two or three inches wide. By comparison, these green crabs were distressingly large, some as big as a human hand. I had a glimmer of what the armored knights of Europe might have felt if, without warning, a squadron of Sherman tanks had rolled onto the battlefield. "That's big," Ruiz said. "Especially when you consider that those little grapsids are the most common crab in the bay—and they're being replaced by these." As alarming as the green crab's size is its rate of growth. In other parts of the world, a juvenile green crab takes two or three years to reach adult size; in Bodega Harbor, it takes only a year. At two months, a juvenile green crab in Bodega Harbor is already as large as an adult grapsid will ever grow. With each generation reaching reproductive maturity so swiftly, the local green crab population has skyrocketed.

Where the green crab has gone, Ruiz has followed: To Bodega Harbor, an ecosystem he began studying as a graduate student at the University of California at Berkeley, years before the green crab first reared its pin-eyed head there. To Chesapeake Bay, where Ruiz established his lab at SERC in 1989. To Tasmania, where, at the invitation of an Australian national marine laboratory in Hobart, he would spend several months establishing field sites and a green-crab research program. Even to Martha's Vineyard in Massachusetts, where one of his graduate-student advisees was completing a doctoral dissertation on the impact of the green crab on the centuries-old shellfish industry. There are many compelling ecological invaders in this shrinking world, more with each passing day, but rarely is an ecologist presented with the chance to study—from time zero or near so—the progress of a single invader in several ecosystems around the world simultaneously. For Ruiz it is an opportunity to continue the slow piloting of invasion ecology from the shallow waters of particulars and anecdotes—the study of one organism in one environment at one point at time—and into a deeper discussion of patterns, tenets, and predictions.

"It's hard to say what any one species can do. That's one of the key questions—and one of the reasons why ballast water is so problematic. It's not at all clear when you release a species how it will perform. Will it become colonized? That question plagues all aspects of invasion biology, from terrestrial invasion to biological control. I mean, you might release an insect for biocontrol and it doesn't establish. You release it a second time and it doesn't establish. And then the third time it does. And then, once an organism becomes colonized, the question is, what is it going to do? We can't answer that, but we can make good guesses. That's why the green crab is such a good model, because it's found across different environments. Critics would say we don't have a good experimental control. But with a wide background of data and convergent types of experiments, we can make strong statements, not only about how a community functions but about how populations respond. The scale at which you gain insight is unparalleled. I guess it's what you'd call a natural experiment. You can do a very elegant lab experiment that's very well controlled, but it's in a lab. You can't transfer it to the field. In the field, things happen that you didn't take into account, and then you learn something about the environment."

Ruiz runs his Bodega Harbor research out of a laboratory room in a concrete building nearby, on a rise overlooking the ocean and a small embayment called Horseshoe Cove. The laboratory belongs to Ted Grosholz, a marine biologist at the University of California at Davis. The two have collaborated on green-crab research for years now. What the biologists Gordon Rodda and Gad Perry are to the study of the brown tree snake, Ruiz and Grosholz are to *Carcinus maenas*: assiduous collectors of data, conducting meticulous experiments and meticulously separating the facts of nature from the artifacts of their experimental design. The field station building houses various Davis researchers. Inside is a labyrinth of low ceilings, narrow corridors, and doorways that open onto cramped laboratory spaces. The air reeks of formaldehyde and stagnant tide pools. It would be a dismal work environment if not for the stellar view: every window opens onto either the Pacific, hazy and bright blue in midmorning, or the dazzling iris of Horseshoe Cove.

When I had first stopped by the Grosholz lab the previous afternoon, I found Ruiz sitting on a rickety swivel stool amid disarray. Two tall rolls of heavy wire mesh were propped up in one corner; a stack of white industrial buckets sat in another. A crumpled bag of corn chips lay on the floor beside the splayed contents of a metal toolbox. Grosholz himself was outdoors behind the lab. With a pair of wire cutters in one hand, he was struggling to transform a third roll of wire mesh into a neat series of small cages. His fair skin was mottled from sun and aggravation; his eyes, an Irish blue, had turned stormy. It was as though I had stumbled onto an otherwise friendly neighbor throttling a groundhog. When he stood up, a curl of mesh sprang up and clapped his leg like a bear trap. "It's amazing as a field biologist how much you learn about shop practices," he said.

Inside, Ruiz, in preparation for an upcoming field experiment, was trying to lasso crabs. A metal tray on a countertop near him contained a dozen small members of a sand-colored shore crab, *Hemigrapsus oregonensis*, that is native to the bay. They scuttled to and fro, their legs clicking across the tin tray; collectively they gave off a sour smell, like cat food. Ruiz held one crab between his thumb and forefinger; he had it by the back of its carapace so that its pincers faced away from him. With his other hand he attempted to work a small loop made of fishing line around the crab's shell and secure it there. It was a delicate task, the crab as intent on attacking the cowboy biologist as Ruiz, his fingers feinting and weaving, was on harnessing the crab. Trials with several different knots and loops had yet to produce one that stayed put. "Different crabs have different carapaces," Ruiz said, "so what works for one doesn't always work for another."

Ruiz has studied the ecological dynamics of Bodega Harbor since 1979, when he chose it as the ideal site for his doctoral research. Grosholz joined the effort in 1993. Over the years, they have developed a solid backdrop of data against which the more recent incursion of the green crab can be studied in relief. Even with all biological introductions aside, Bodega Harbor experiences countless fluctuations. If the water warms up later in the spring than usual, the native shore crabs get a late start on their reproductive season, so their numbers are below average. With fewer crabs to feed on, the shorebirds focus their appetites on other invertebrates, with the result that the following year the clam population is smaller than usual. And so on—ripples of flux playing back and forth, crossing, enforcing, and damping one another. Suddenly the European green crab arrives, like a stone tossed into a pond. What is its impact? Since its arrival, Ruiz and Grosholz have noted that the population of shore crabs and of two pearl-size native clams have plummeted. But there are other potential factors. Thirteen local species of shorebirds—plovers, godwits, turnstones—are known to prey on marine invertebrates; perhaps they figure in the decline. Perhaps the green crab merely amplified a wave of change already under way. Circumstantially, the green crab is the culprit. Ruiz and Grosholz would like to rule factors out or rule them in, in proper proportion.

"We're trying to really test the hypothesis that the green crab is actually having an impact," Ruiz said. "A hundred and one things are changing simultaneously; it's very difficult to tease apart those changes. We need to make an accurate case. One of the big questions is, what are the consequences? With all the hoopla about introduced species, it seems like there must be some clearer examples. The more I'm in the business, the more I realize those questions haven't been answered." Grosholz, when he returned from outside, added, "I've gone to a lot of introduced-species meetings. There's a lot of hype, a lot of prediction. But you have to be careful. Once you oversell it, your credibility is lost. We need to have a few good examples in hand in which the data are good and solid. That's what we're doing here—providing one of those examples."

A simply stated hypothesis—the introduced green crab is depressing the population of native shore crabs in Bodega Harbor—in fact disguises a host of assumptions, all of which needed to be tested and confirmed before reaching any general conclusion. To start, do green crabs even eat shore crabs? If so, how many do they eat—and is that number significantly larger than the quantity consumed by shorebirds? Is it enough to lower the population of shore crabs? How quickly do shore crabs and green crabs reproduce and grow? So Ruiz and Grosholz spend their time in Bodega Harbor running experiments: designing and devising them, setting them up, conducting them, repeating them, tabulating and deciphering the results, teaching junior field assistants how to run them in their absence, and, often as not, starting again from scratch. To better discern how heavily the green crab preys on the native grapsid crabs, one experiment would examine how heavily the grapsids are preyed upon by everything else in the bay. The crab in Ruiz's grip—after several attempts Ruiz had succeeded in securing a loop of fishing line around it—was a designated volunteer in this experiment. Ten shore crabs would be tethered one by one to ten railroad spikes sunk in a row into the mudflat at low tide. When the tide rose, the crab would be free to roam within the one-foot radius allowed by its leash, but it would otherwise be at the mercy of all passing predators, like a submarine Prometheus. Ruiz now threaded a fishhook onto the crab's harness; he fixed the hook in place so that it curled, bitter end up, from the back of the crab's carapace. Any fish that attempted to eat the crab would be caught, providing the scientist with some evidence of the identity of the crab's non–green-crab predators. "Basically, we're fishing," Ruiz said. He held the crab at arm's length to assess his handiwork. The crab expressed a faceless, waving fury. "I wouldn't say he's happy, but he looks okay to me."

Another experiment aimed to characterize the green crab's appetite.

This involved a series of large plastic jars, each one containing several inches of mud from the bay, one green crab, and forty-five small clams. Left alone with the clams, how many would a green crab eat in a day? Ruiz emptied a jar into a metal sieve in the sink. The wet sediment drained away, exposing a flustered green crab and a hash of shell rubble and pearl-size clams. Ruiz plucked out the surviving clams one by one and made a note of how many had been eaten by the crab. For decent results, the experiment would have to be repeated several times: hours of sifting through shell shrapnel, counting fatalities, keeping score. I may have rolled my eyes.

"This is what most research is like," Ruiz said. "It's not just driving around San Francisco Bay pulling up buoys, drinking coffee in interesting shops. After the first time, it's repetitious, getting enough data to convince people that you know what you're talking about."

The clam-eating experiment was mere prelude to the main event, which would take place outdoors. Ruiz and Grosholz would set up an array of small cages on the mudflat, then place in them differing pairs of crabs: one green and grapsid; or two greens; or one of each, of varying sizes. The crabs would be left alone under the rising tide, and a day later, the researchers would return to count the winners and losers, or the remaining shreds thereof. The experiment would be repeated numerous times in order to clearly establish an answer to one question: If a green crab and a hemigrapsus shore crab are left alone in the wild, or in some experimental approximation of it, will the green crab eat the shore crab?

"It's a necessary thing to demonstrate if we want to conclude that crabs are declining because the green crab is eating them," Ruiz said. "One advantage of doing it in the field is that the green crabs have alternative food. It's not like being thrown into a situation like the Donner Party, where they eat their neighbor or they die. It can give them an opportunity to eat other things if they wish. But if they want to eat one of these hemigrapsids, so be it." All of these experiments would require green crabs, however, so at dawn the next morning Ruiz set out across the subtidal plain of Bodega Harbor, his boots slurping in the mud, a white plastic bucket swinging from his hand, to collect volunteers. If the beady eyes of a green crab had been sharp enough, they would have detected the crouching forms of Grosholz and a junior assistant a hundred yards out on the mudflat, setting up the day's experiment. But as natural selection had worked it, all a crab volunteer could discern was Ruiz's black rubber boots and the grasping hand of science plucking it from under its seaweed shelter and dropping it unceremoniously into the bucket.

One advantage of having ten legs is that it eases one's progress across mud. For a biped, however, walking on mudflats can be treacherous business, not far short of braving quicksand. For every mudflat Ruiz has trod upon, he has met a marine biologist with an allegedly foolproof scheme for remaining above it. One colleague, taking what Ruiz calls the Zodiac approach, designed a pair of boots that inflate like rubber Zodiac dinghies and, when worn, distribute his weight over a wide surface area. Another colleague ties one end of a rope around his waist and the other to a tree; if all else fails, he can pull himself back to shore hand over hand. At the recommendation of another biologist, Ruiz once tried wearing snowshoes. He made progress for fifty yards or so, until the mud became so soft that the shoes sank below the surface, at which point the snowshoes became anchors and Ruiz could not move his feet at all. On another outing, a colleague of Ruiz's brought along two pairs of water skis, one for Ruiz and one for himself, to strap to their boots. Twenty yards out, the colleague caught the edge of a ski and fell forward into the mud. Even now, Ruiz roared with laughter as he told the story. "He couldn't get back up!" The colleague, exasperated and unable to move forward, finally removed the skis altogether. Then, life-preserver style, he threw one ski ahead of him, dragged himself up behind it, threw it again, all the way back to terra firma.

I kept pace with Ruiz as best I could. A pair of vultures disguised as seagulls trailed us in the air. From the direction of the mouth of the bay, a mile or so in the distance, came the moan of a foghorn and the rhythmic, honking bark of sea lions. "They're shark food," Ruiz joked. "They're saying, 'Eat me, eat me!'" Ruiz has developed a rigorous method of measuring the abundance of green crabs in the bay: at low tide he walks far out on the mudflat and fixes a string several meters long—a transect—to the sediment; at high tide he returns and swims along the transect in snorkel and scuba suit and counts the crabs in that standard distance. It has not escaped his notice that a marine biologist in a scuba suit looks dangerously like a sea lion. For the moment, however, walking on the emptied tidal flat, he contented himself with the guesstimate method of assessing green-crab abundance. "Here's a *Carcinus!*" He held up a green crab he had nabbed from the matting of seaweed underfoot. It was a juvenile, its carapace little more than an inch wide, and it was a translucent green color, as if it had recently molted. It was an angry youth: the animal could not reach Ruiz's fingers, which held it from behind by the carapace, so it made a blustery show of shredding the air with its tiny pincers. "We've come across three in a square meter. That's the densest we've seen so far." Ruiz continued rummaging through the algae underfoot. He pointed out a white sea slug the size of a thumb—an opistobranch that had arrived from New Zealand in the early 1990s. He found three more green crabs, which he dropped into the bucket, and a small, native Dungeness crab. My eye was drawn to another sea slug, this one native, meandering among the strands of seaweed. It was transparent, with two neon-blue stripes down its length and a mane of tendrils, the sea slug version of lungs, rising from its back.

On gathering a few dozen green crabs, Ruiz turned around and trudged back in the direction of the laboratory building. I continued on across the mudflat toward Grosholz's crouching form. When I reached him, he was busy pressing crab cages into the sand. Each cage was roughly one cubic foot in size and made of the heavy mesh I'd seen him battling the day before. The top end had a lid attached by plastic cable ties. The bottom end was open. Grosholz placed this end on the wet sand and pushed down on the cage to set it several inches deep and firm it against the tide. It was heavy work, and Grosholz's face had again taken on a mottled cast. Twenty cages would be arrayed a meter apart in a small grid. By eight-thirty in the morning, a third of the cages had been set up and a rough suburb of streets and avenues had taken form. Ruiz would return later with crabs to place inside. A second, larger tract of cages sat in the sand nearby. They had been erected the previous day, and the intervening tides had pulled some cages slightly off axis and plastered others with seaweed, such that the array resembled a miniature fishing village in the wake of a hurricane. Far across the tidal flat, I could make out a large skeletal wooden enclosure that rose from the water in the channel at the entrance of the bay. Grosholz referred to it as the Monstrosity. It had been erected several seasons earlier by a scientist studying the feeding behavior of sharks; construction had halted when the grant money ran out. The half-finished structure loomed now above passing trawlers, looking positively Ozymandian.

Like Ruiz, Grosholz entered the field of ecological invasion sidelong from zoology. He studied basic biology as an undergraduate; after college, in the early 1980s, he took a job in the Caribbean helping to improve the local mariculture of conch, a staple meat in that part of the world—"the roast beef of the sea," Grosholz calls it. He was a sea farmer. A couple of years of that, and he was already hitting his head on the professional ceiling. He returned to the States and enrolled in a doctoral program in zoology at the University of California at Berkeley. There he met Ruiz and, having met Jim Carlton—"the founding father"—some years earlier, he was soon pulled into the inner circle of marine bioinvasion. Invasions had become academically hot; the odds of winning funding for a study of ecological impacts or ecosystem-wide dynamics improved if you could justifiably add the words *invasion* or *introduced species* to your proposal. Before long, Ruiz had coauthored as many papers with Grosholz as he had with Carlton, many of them derived from their fieldwork and experiments together in Bodega Harbor. At aquatic-nuisance-species conferences, the two scientists can often be found in some corner of a hallway, huddled in conversation. Their families vacation together.

"We've worked together so much, we bicker like old men," Grosholz said. "That's partly what happens when you have two people who are used to running their own shows. Greg runs a lab of eight or nine people, and he's the director. And I've got my own smaller show. So when we get together, it's 'Do it my way!' 'No, do it my way!' It's good: we catch ourselves making mistakes, or we provide a perspective that we wouldn't otherwise have. It's easy to convince yourself you've thought of everything. Academia promotes a kind of lone-dog mentality; that's part of the training. It's different from a business, where everyone's working like a team. It helps to be friends, because then you can just say, 'Oh well.'"

If Ruiz and Grosholz have inherited one lesson from Carlton, it is an appreciation for the true difficulty of extracting information from nature. The two are joined in a tenacious pursuit of hard data. Though they often disagree on the finer ecumenical points, they share a deep and abiding faith in the revelations of statistical robustness—that is to say, repeated repetition. You build cages in the mud, you put scissor-handed crustaceans in them, and then you stick your fingers in again and again and again. Grosholz's assistant, a Davis postdoc named Kim Shirley, was doing exactly that now with the previous day's experiment. While Grosholz pressed new cages into the flat, Shirley worked her way down the seaweed-draped avenue of old ones with a pair of heavy clippers, snipping the cable ties that held the lids closed. The setup was designed to see whether green crabs prey on shore crabs as readily in the wild as they do in the laboratory. Each cage held either an introduced green crab and a native shore crab, or one of either, alone, as controls. One by one Shirley peeled back the cage lids and rooted around in the sediment with her hand to tally the survivors. She retrieved feisty green crabs, shore crabs, or the disembodied legs and pincers of the latter. Often, however, she found nothing at all. Evidently the volunteers had escaped through a newly apparent flaw in the experimental design—perhaps through gaps where a cage had not been sunk deeply enough into the sediment, or, less likely, through the small windows of mesh. Some cages held new volunteers, crabs that had crawled in after the experiment began. Here was an epistemological dilemma. Unless the absence of every shore crab could be confidently explained—either the crab had definitely escaped or it definitely had been eaten—the data would reveal nothing about the feeding behavior of green crabs and everything about poor construction.

Grosholz sighed. "Anytime you put a cage out in a natural system, there are all sorts of experimental artifacts. So there's a suite of stupid technical questions about how to make a cage work, mixed with real questions about the biology of the animals." He renewed his effort to sink the new cages deep into the sediment, to prevent any further escapes. The sun had risen sufficiently to compel him to remove a fleece jacket he had worn since dawn. The fog had burned off, revealing a crystalline blue sky. At around ten, Ruiz sauntered out from the lab carrying a bucket of crabs. The scratch and scrabble of a few hundred crustacean legs was audible from a dozen yards off. To ward off the sun, Ruiz had found a camouflage cap, which he wore now at a jaunty angle. He watched quizzically as Grosholz wrestled a cage into the sand.

"You're like five centimeters down there," Ruiz said. "We don't need to go that deep." He floated the idea of building the cages out of hardware with an even smaller mesh size, to prevent possible escapes. It would be stiffer than the current material and harder to bend into cages. Ruiz added, "I guess it's a bad time to bring this up." The prospect added a new shade of irritation to Grosholz's face.

"It'll be a mess to work with," Grosholz said.

Several yards away, Shirley continued checking the old cages for crabs or former evidence of them. "If there's an interaction, you'll find some part of a crab," she said. In one cage, she found a green crab and the pincer of a shore crab. In another, a large, snapping green crab and, wedged into a corner, a shore crab—alive but lacking three legs. I wondered if the ancient Roman Colosseum had had janitors and what that job was like. To Shirley I said that the term *interaction* seemed like an understatement.

"You mean between Greg and Ted?"

Sure enough, Ruiz and Grosholz had found a new point of disagreement. Their voices floated above the emptied tide bed.

"I'm not disagreeing with you."

"I'm just saying these cages are so old, we should just get rid of them."

"I'd just like to salvage some of these data."

For Ruiz, every reliable data point is another solid dot in the line of dots through which he hopes to connect invasion rates—the amount of ballast water globally in motion and the number of invasions occurring from region to region—with actual invasion impacts. What would be ideal, he thinks, is if invasion biologists could all agree on the kind of dots to use. Once, he invited a large number of marine biologists to SERC—Jim Carlton and Andy Cohen came, as did marine biologists from Hawaii, Guam, and elsewhere around the country and world—to discuss future avenues of invasions research. Since 1995, when Carlton and Cohen published the results of their survey on the nonindigenous inhabitants of San Francisco Bay, other marine biologists have rushed to count the introduced species in their own neck of the sea. All fine and well, Ruiz said, but there was a need for standardization. If varying scientists survey varying environments using varying sampling techniques over varying timescales, it becomes difficult to assemble the data into a single, meaningful global picture. And a global picture would be useful, both to inform the rafts of ballast-water policy being cobbled together, and to probe some long-standing ecological theories—about propagule supply, about larval dispersal, about community structure. Innumerable, replicable, golden bricks of data pave the road leading from here to a unified invasion science—capital I, capital S. "I think what you're seeing here today is the struggle to become that," one participant said afterward, impressed. "Whether it happens now or in ten years, whether it's these people or some other people, I don't know. It will have a real influence on ecology. Some of my colleagues in ecology haven't been able to see it that way; they think this is just about zebra mussels."

Grosholz has a phrase for the revolution under way: *Turning MacArthur on his head.* It is a reference to the late ecologist Robert MacArthur, who in the 1960s, with Edward O. Wilson, developed the theory of island biogeography, a sweeping reassesment of how nature works. It grew from their effort to explain why islands have fewer species than continental areas of the same size. Their answer was a remarkably straightforward mathematical equation: the number of species at any given time reflects a balance of the number of species immigrating to the island versus the number going extinct. Picture two lines on a graph plotted over time, one (the immigration rate) sloping down, the other (the extinction rate) sloping up. Now imagine some species making their way to an empty island over geological time—spiders ballooning in, a finch blown adrift, whatever, however. Over time, as species arrive, MacArthur and Wilson wrote, the immigration rate falls; there simply isn't room or nutritional support for everybody. Likewise, over time, the extinction rate rises as more species arrive to potentially become extinct. The point where those two lines cross, incoming versus outgoing, represents the present, the number of species on the island right now. That number will be bigger on larger islands, both because a larger area will catch and support more immigrants and because species-extinction rates are generally higher on smaller islands. (Species on small islands tend to have fewer members, so are more prone to go extinct in one fateful blow.) The number of species will also be larger on islands closer to the mainland, as near-shore islands have higher immigration rates than remote islands: statistically, they're easier for an immigrant to reach.

In short, the theory proposes, the number of species on an island varies directly with the island's size and inversely with its distance from the mainland. In many ways it was a radical concept. The prospect that nature, with its infinite variations and subtle interactions between living things, could be expressed in mathematical, almost mechanistic terms was tantalizing. It suggested more broadly that biological diversity, the number of species in an area, could be considered strictly as an outgrowth of physical aspects of the landmass. Set aside niches, food webs, competition between species, rates of population increase, all the dynamics traditionally thought to bear on how ecosystems gain form. Chance—the random arrival of a species of whatever kind; its characteristics were mathematically irrelevant—could be just as important as the usual "assembly rules," perhaps more so. Finally, the theory codified the idea of nature undergoing perpetual change. The rates of immigration and extinction met at a point, an equilibrium number of species that would remain constant over time. But it was an equilibrium in number only: the taxonomic makeup of that number—the kinds of species actually involved—could vary constantly, from day to day or eon to eon.

Or so it seems now to some biologists, in retrospect and on paper. Since its inception, the theory of island biogeography has been a magnet for discussion, argument, experimenting, and further theorizing by ecologists eager to test, probe, confirm, or deflate it. Ruiz and Grosholz are no exception. In their view and that of several colleagues, ballast-water invasions offer an opportunity to reexamine and perhaps recast a central tenet of the theory. MacArthur proposed that as the distance to an island increases, the rate at which new species immigrate to it decreases. That may have been true in the distant past, before the travel of planes and ships added a secondary layer of faster-moving immigrants. But in the modern world of shrinking distances, MacArthur's equation wants the addition of an alternate variable, one that takes into account the likelihood of a new species arriving by human device. For shipping, that variable might be expressed as the total amount of ballast water arriving annually in any one harbor, or the potency of any one ballast load, or any of several similar factors. When Ruiz says, casually, "I like the idea of looking at islands in terms of vector strength," what he means is, Let's look at MacArthur from another angle.

To an invasion biologist, MacArthur may require inversion for another reason as well. Interpreted at its purest, the theory of island biogeography forswears any notion of a "balance of nature." Immigration and extinction might balance out to create a constant *number* of species over time, but the membership of that ever-constant diversity is perpetually turning over: any given species might be here today and gone tomorrow, for any reason. Moreover, that number is shaped strictly by the size and remoteness of the island, not by any ecological goings-on between species. Consequently the rate of extinction on an island should be wholly unrelated to the rate at which new species arrive. At time zero, the extinction rate is zero because no species are yet present to go extinct. Extinctions rise with time because more species, having immigrated and being present, are on hand to become extinct. It's math, nothing more.

However, the theory's authors were not content to leave it at that. "It is not clear whether MacArthur and Wilson fully appreciated the implications of this radical assumption," the ecologist Stephen Hubbell writes. They invoked competition, the role of niches, the varying colonization strategies among species, the pressures of coexistence. As a result, through the filtering effects of a generation of biologists, the theory of island biogeography has come to picture nature as a kind of delicately balanced machine, not so unlike what Elton envisioned. The rate of immigration declines over time owing to the increasing pressures of competition. The rate of extinction rises as new species arrive and crowd old ones out. These are precisely the ideas that ecologists have been so hard-pressed to demonstrate in the laboratory and the field. "The number of cases in which local extinction can be definitively correlated to competitive exclusion is vanishingly small," notes Hubbell. The study of invasions, both terrestrial and marine, has cast added doubt. The number of species in San Francisco Bay and other bodies of water around the world has not remained constant over the past century, but has risen markedly. Competition does not appear to be keeping new species out.

"The issue is the MacArthurian nature of the world," Grosholz says. "The MacArthurian world is a tightly structured world without much space for new competitors. What invasions have shown is that there are plenty of unused resources. Ecosystems can absorb a lot of new species. I mean, holy cow, look at San Francisco Bay with two hundred–odd nonindigenous species! Who would have thought an ecosystem had that much unused niche space?"

In effect, biologists have spent the last several decades developing powerful mathematical and statistical tools with which to quantify the rules of nature, only to discover that the rules are far more plastic than previously believed. The revelation is both exciting and unnerving. Nature can absorb a lot. Between the prelapsarian world—the state of nature as it was prior to the arrival of humans—and the modern one infested with flying snakes and continent-hopping crabs, there is a great deal of give. The changes are more subtle, the stakes more difficult to discern, the losses harder to tabulate.

The paradox of biological invasion, and one reason scientists have such difficulty articulating its hazards, is that to the average backyard viewer, the gross result appears to be ecological addition rather than subtraction. Some invasions, like that of the brown tree snake, do indeed cause a direct and dramatic removal of native species from an ecosystem. But such cases are by far the exception. Most successful invaders simply blend into the ecological woodwork; some may cause perturbations—the mynah bird that abets the spread of weedy plants; the marine pill bug that eats the roots of mangrove forests—but often the main impact is simply an increase by one of the number of species in the ecosystem. To the local eye, biological diversity seems to have increased. Isn't that a good thing?

Here an ecologist steps back to distinguish between "alpha" diversity and "beta" diversity. Alpha diversity is the number of species in any given location, A, B, or C; beta diversity represents the relative diversity between those locations. If a New York snail invades San Francisco Bay, and a San Francisco snail invades New York Harbor, the alpha diversity in both locations has increased, but the beta diversity has decreased, as the two environments now share two species; each place is that much less unique. In effect, beta diversity is the fancy measure of homogenization, and one can see why its plight is hard to impress on the public. Most people live small, local lives and are grateful for whatever manages to thrive in their arena; they live in an alpha-diversity world. Whereas beta diversity is visible only on a grand scale, requiring some effort to take in; it speaks to the traveler and the reader of travel books. Its appreciation is a kind of luxury, although perhaps no less valuable for being one, the traveler would say. Grosholz would say that its appreciation points up the risk of valuing nature strictly by the numbers. A head count of species—native or introduced—in any given place is one way to measure the impact of biological invasion, but it may not be the most telling.

"The key question is, what is the impact? What effect does it have? Does it matter? Extinction may not be the only issue. That's the main difference between marine and terrestrial ecosystems. With the brown tree snake you can point to species and say, 'Look, those things are gone.' With marine species it's not so easy; you may not see actual extinctions. You can get qualitative shifts in communities if a species falls below a certain population threshold. We may have to focus on that level of analysis. There may be extinctions, but other effects could emerge that are significant on an ecosystem level. I'm more concerned about those kinds of changes. We should focus on ecosystem management, not just species management. Extinction is a warning sign, but equally important are fundamental changes in ecosystem structure. Where do we draw the line? Maybe we have to say, 'We care above this line, and we don't care below that one.'"

One day back at SERC, I sat with Ruiz on a picnic table on the shore of a Chesapeake inlet, eating sandwiches we'd picked up in town. The air sang with insects and the vitality of late spring. A small fleet of Boston Whalers, used for research, was moored to a dock that floated on bright green water. Ruiz pulled his sandaled feet onto the table bench and sat cross-legged. Even a quarter mile from his office, he seemed to have shed a layer of formality, and the indoor Ruiz and the outdoor Ruiz had achieved a moment of equilibrium, however transitory.

"In the sixties and seventies, the goal in ecology was to look for pristine ecosystems," he said. "But over the course of ten years it has become clear that there's no such thing. I think the most useful thing we can do is point out that we don't live in a pristine environment, but that we are having an enormous impact. And to point out what those impacts are, so that we can make decisions about what we want."

I asked, "Are invasions bad?"

"I'm not saying they're bad. The pattern is interesting. The solution is, where do you start to draw distinctions? That's a social dialogue. How much do you care? There should be no illusion that we live in a pristine environment, or that there is such a thing. So then the question is, can we accept that level of change? What's acceptable to one person is not acceptable to another." A black garter snake had chosen that moment to slide out of the grass and across the gravel driveway, from one patch of greenness toward another.

I asked, "And what's acceptable to you?"

"I try not to answer that, because it puts me on one side or another. I do think the number and extent of invasions is alarming. But I don't try to hype it. Our role here is to provide information to clarify the risks. Professionally, that's where we are. Personally, what's acceptable to me is to have an explicit dialogue about what those risks are and what we consider acceptable. I may not agree with the outcome, but I do feel there ought to be a process that allows for a dialogue."

22

In the afternoons and evenings, when Murphy and Steves were not on deck lugging equipment from tank hatch to tank hatch and pumping the SeaRiver *Benicia*'s ballast water into sample bottles, they were belowdecks in a small room that served as their staging area, unpacking equipment and making preparations for the next day's sampling effort. Already they had collected nearly two dozen gallon jugs worth of water, which had been stacked in the shower stall to form a semiaquatic pyramid.

Increasingly during these hours I gravitated upstairs to the wheelhouse and bridge. There was something hypnotic about the perspective from up there. The view was better, certainly, especially late in the day, when a thin layer of mist formed just above the sea surface and the dropping sun infused it with an amber glow. Also, the perimeter guardrail around the deck no longer loomed as large from that height, which I found comforting. We were a block in the stream, and I was a few dozen yards up, looking down on it. To my eye, the deck rail offered about as much security as any barrier reasonably could in the circumstances, but of course it could not possibly impede someone careless or determined enough to fall overboard. It is remarkable how most preventative lines drawn by humans on the landscape—fences, barriers, borders—provide more mental than physical protection. The central stripe on a highway will not prevent traffic in the opposite lane from swerving head-on into this one, nor vice versa. The line works solely because we obey it. Safety resides not in the line per se, but in the commitment to draw it.

An organism in the sea lives sandwiched between two parallel and impassable lines: the sea surface above and the seafloor below. As on land, the eventual trajectory is downward. Even the tiniest plankton are engaged in an ongoing effort to defy gravity, to maintain their altitude as long as possible, to slow the inevitable sink. Accordingly, their bodies are generally small, often microscopic, and composed almost entirely of seawater. Their slight masses are spread thin, such that the density of an organism is hardly greater than the medium it travels in. Their surface areas are large relative to their volumes: greater resistance against the water means a slower descent. The dinoflagellates and diatoms, single cells of phytoplankton, may be elongate, or they may bristle with spines or horns or sport a glider's wings. Others may be linked in chains that trace slow downward spirals through the photic zone or zigzag in their fall like autumn leaves. The largest zooplankton are bare ribbons of living matter; the smallest possess long, feathery appendages that trap food and, incidentally, increase their surface area. I followed these thoughts all too often as I lay awake in my cabin listening to the ship churn through another night of miraculous mechanical suspension on the sea. Life is a choice that most organisms have no choice but to make in the positive. It is a delay game; the end is resisted and slowed because it must be. How else is one to live?

Among its amenities the SeaRiver *Benicia* had a small exercise room. For want of alternatives, I took to working out on the rowing machine. It was half the workout I had hoped for, as half the time the steady roll of the ship had me effortlessly rowing my skeletal boat downhill. I have since read that copepods, the microscopic oar-footed plankton so common to the seas and to ballast tanks, in fact are capable of propelling themselves tremendous distances up and down through the water column. One species can ascend fifty meters in an hour, roughly equivalent in human scale to the pace at which one might walk to the corner deli. Indeed, whole clouds of copepods and other zooplankton daily undertake great vertical migrations in pursuit of dinner. What is the aim of this persistent venture? Where are they going and why? I was told of a sixty-year-old man, an occasional crew member on the SeaRiver *Benicia*, who puts in ten thousand meters a day on the ship's rowing machine. Evidently he is attempting the stationary equivalent of a row around the world; at his current pace, he will be rowing until the age of seventy-two. In 1999 Tori Murden became the first woman to cross the Atlantic alone in a rowboat. Her first attempt failed; she injured a shoulder during a hurricane and was forced to abandon her boat. Two months after her rescue, Kevin Duschesnek, the third mate aboard a SeaRiver tanker bound for France, spotted her boat drifting off the coast of Portugal. He notified the captain, who changed course, picked up the boat, and shipped it back to Murden.

Duschesnek is third mate on this voyage of the SeaRiver *Benicia.* He takes the helm after midnight, when nothing at all can be seen. After nightfall, darkness shrouds the wheelhouse, and the visible horizon shrinks to the distance of the windowpanes. The room itself is kept dark to ease the reading of navigational instruments, the celestial guidance system, and the radar, which can be tuned sensitively enough to pick up breaking whitecaps. The world, at least to the eye of a traveler not involved in guiding the ship, is reduced to a blue glow and its inward reflection on the wheelhouse windows. The helm itself is a small steel wheel on a panel console. The ship is steered according to a digital compass readout on the wall that shows the ship's heading in degrees from 0 to 360. We are bound for the number 328, roughly north-northwest. The only physical bearings are lines on paper, the ship's course marked across a navigation chart. According to the chart's hydrographic lines, we are passing over a stretch of ocean floor called the Alaska Plain that lies ten thousand feet below. Here and there, the snuffed fuses of extinct and sunken volcanoes rise high above the seabed, though they crest two to three thousand feet beneath us. Durgin Seamount, Applequist Seamount, the mammoth peaks of the entirely submerged Gulf of Alaska seamount province. Another chart, of an area nearer the Aleutians, might show Meiji Seamount, the primordial start of the Hawaiian archipelago, eighty million years old and a mere hillock now, five hundred fathoms down. The existing chart showed, still far from us to the northwest, the trail of Captain Cook, who came to Alaska after visiting Hawaii and named large portions of the landscape for himself or his peers—Cook Inlet, Vancouver Island, Prince William Sound.

The act of discovery, it seems, is incomplete without names. To name a wild space is to tame it slightly, smooth its rough edges, yoke it nominally to the human realm. It is no longer other; it is us. The name might be descriptive of a prominent feature of the terrain (Owyhee, "the hot place"; Honolulu, "fair harbor"; Edgewater) or utilitarian, highlighting the advantages of the location or the profit to be gained there (Silverton, Pie Town). A place might be named for the person who found it, or friends of the discoverer, or benefactors or saints. Some place-names express the hope that the place will take on the attributes implied by the name and so become doubly familiar. This may explain the prevalence of shared names around the world: Inverness (in Scotland, Nova Scotia, northern California), New England, New Amsterdam, New Holland,

New York. According to *The Story Key to Geographic Names*, Scotland takes its name from the Celtic word *scuit*, meaning "wanderers" or "fugitives"—it is a land of exiles and travelers, or was until these people discovered the place and settled on a name for it. Wales is from the Anglo-Saxon *wealas*, or "foreigners." Great Britain, this book contends, is a combination of the Celtic *bro*, or "region," and the Basque *etan*, meaning "those who are in." In effect it means "the people who are in that region"—either "here" or "there," depending on whether you are within or without its boundaries. In the mid-1800s, when much of New York State was still being mapped, Robert Harpur, a clerk in the land commissioner's office in New York City, was given the task of applying names to a growing number of charted dots where towns were sprouting up. Harpur reached into his own history as a classics scholar and sprinkled central New York, where I grew up, with ancient greatness: Syracuse, Rome, Troy, Ithaca. Wherever we live, we are all living somewhere else. Without names, words, lines, we would be lost.

With the third mate at the helm, the SeaRiver *Benicia* continues onward toward 328. Compared to the steering wheel of a car or even a yacht, the helm of a ship this size provides little immediate physical feedback. It turns smoothly without resistance or effort, quivers not the slightest even if the rudder is shuddering in high seas. All progress is abstract, any direction purely conceptual. Since the sixteenth century, firm believers have insisted on the existence of a vast hole at the North Pole that enters a hollow Earth and exits at the south. A Utah company advertises a trip aboard a Russian icebreaker to enter the North Polar Opening (84.4° north latitude, 141° east longitude) and visit the Inner Continent, where expedition members can ride a monorail to the lost Garden of Eden. I imagine an experience more akin to the voyage of the SeaRiver *Benicia*. By day, we roll like a giant Buick across a desert sea. In darkness, we are a slowly swaying elevator, a capsule floating in a nameless ether. We are no longer nowhere: now there is only here. If and when we should emerge again into light, it will come as a surprise to discover that there is anywhere else at all.

23

In late January of 1777 Captain James Cook, having rounded the Cape of Good Hope and run eastward before the forceful winds of the southern fortieth latitude for several weeks, anchored in a protected cove along the southern shore of Tasmania; he named it Adventure Bay. His homeport, the town of Whitby in northern England, was six months in his wake. This was his third and final voyage. He was bound for New Zealand and afterward Tahiti, where he intended to set loose a bull, two cows, and several sheep he carried on board the *Resolution*. Soon after that, he would discover Hawaii, or at least place its people on European maps; sail on to the Pacific Northwest to search in vain for a Northwest Passage; name an Alaskan inlet for himself at the entry point to present-day Anchorage; trace the outward line of the Aleutian Islands; and then make his way back to Hawaii for an unexpected, violent final rest.

But all that was months away. At the moment, in Tasmania, the future cattle of Tahiti were hungry, so Cook and his crew rowed ashore to gather grass and to cut wood for the ship. They tied their boat to a tree; Cook climbed another tree for a better look at the isthmus. Before long, they were approached by a band of Aboriginal men. In his log, Cook describes them as confident and entirely naked, with dirty teeth, woolly hair sticky with red ointment, and dark skin abraded and raised in ornamental lines and waves. They carried one weapon among them, a pointed stick. Cook offered them fish and loaves of bread; the men sniffed the items, tossed away the bread, and handed back the fish. Cook wrote in his journal: "They received every thing we gave them without the least appearance of satisfaction." History has been more kind. The tree that Cook tied his boat to still stands today and is named in his honor. Nearby is a monument, a small memorial chimney of white brick. There is also a Captain James Cook Memorial Caravan Park nearby, and a Bligh Museum, named for the infamous Captain Bligh, who passed through in 1777 as a lieutenant on Cook's voyage and again as his own captain in 1788, when he planted Tasmania's first apple trees. Otherwise, the area feels nearly as remote and unpopulated as it must have in that earlier era. The Bligh Museum was closed on the day I visited, however, so after a brief look around, I turned and drove back in the direction I'd come.

Tasmania is a province of Australia, a Pennsylvania-size island off the country's southern coast. Discovered by the Dutch in 1642, Tasmania became known among sailors on the Roaring Forties as the first significant landfall east of southern Africa; it was a fine place to beach your ship, invert it, and scrape off the barnacles and layers of growth that had accumulated there during the weeks and months of sailing. In the nineteenth century, as part of a new British colony, it became a fine place for England to scrape off its overflowing prison population, and Tasmania soon developed a reputation as the world's most notorious penal colony. The last of Tasmania's Aboriginal people, who were otherwise hunted to extinction, died of natural causes in 1876. The province harbors the largest portion of untrammeled wilderness in Australia and manages, in a small frame, to encompass an Earth's worth of topography: temperate rain forests, snowy peaks, turquoise seas, empty white-sand beaches. Its population of half a million is concentrated mainly in Hobart, an unexpectedly lively city of hills and harbors and smart Edwardian homes.

If one drilled a tunnel through the earth downward from Manhattan (being careful to avoid the Garden of Eden), it would open not far from Tasmania. Nowhere have I felt so antipodal: upended, reversed, simultaneously at home and far from it. The month of March marks the onset of autumn; the markets brimmed with varietal apples I had never heard of. The poplars were fountains of yellow. The constellations I could recognize were upside down. Some animals to look out for: bilby, bandicoot, echidna, platypus, Tasmanian devil, and several very poisonous snakes. Wallabies are as common as deer; rural highways are battlefields of roadkill. It was the geographical incarnation of déjà vu. Some days I thought Tasmania was England but with few people. Some days I thought Tasmania was Nova Scotia, a rugged nowhere island, except wombats are even more strange-looking than moose. Some days the fragrance of Tasmania's eucalyptus forests reminded me of California, except California's eucalyptuses were originally imported from Australia. Hobart might be San Francisco, except everyone drives on the opposite side. Some days, after all day looking the wrong way before crossing Tasmanian streets, I forgot which hand is left and which is right, where I'm going next, why.

I had come at the invitation of Greg Ruiz, who had been invited by an Australian marine biology lab in Hobart—at the time, it was called the Centre for Research on Introduced Marine Pests—to come down for several months and continue his field studies of the European green crab, which had appeared in Tasmania only a couple of years earlier, uninvited. When I caught up with Ruiz one morning at the lab, a modern laboratory complex on an updated nineteenth-century waterfront, he was rushing around, preparing for a few days of fieldwork in Falmouth, a small seaside town several hours north of Hobart. He crammed a wet suit and a pair of waders into a canvas bag. "It's beautiful up in Falmouth," he said. "If you can get hold of a car, you should drive up."

The next day, I did. For a headquarters Ruiz had rented a beach house from an elderly woman; it sat on a rise overlooking the shore and was referred to as Miss Lyle's cottage. From his Edgewater lab, Ruiz had brought along a couple of younger biologists including Bill Walton, a graduate student whose doctoral work examined the impact of the green crab on Martha's Vineyard. Kate Murphy, an Australian native, was working at the lab in Hobart. Ted Grosholz was visiting for a few weeks; Jim Carlton promised to pass through but never did. The field site—a tidal lagoon, set back from the shore behind low dunes, that filled and emptied through a narrow, fast-moving stream—was a two-minute walk from the cottage. On the sunny days, as nearly all days were that season, the surface of the lagoon shimmered and turned sky blue. The water was clear and breathtakingly cold. It was the kind of field site to make one wonder why anyone would aspire to be anything other than a marine biologist.

On the morning after my arrival, Ruiz was providing a tour of the lagoon for two marine biologists visiting from the States. They wore sleeveless wet suits and strolled through the shallows, brightly splashing water. Ruiz walked along the shore in a pair of green waders, with one hand towing a small aluminum rowboat. The boat held several rectangle mesh cages designed to catch green crabs. The Falmouth lagoon bore several important similarities to Bodega Harbor that made it a useful point of scientific comparison. The green crab had appeared in Falmouth relatively recently, as it had in Bodega Harbor, so the response of two native ecosystems to a single invader could be studied essentially from the beginning in both places. And the ecosystems themselves were similar: the dominant native predator in the Falmouth lagoon was a small crab, *Paragrapsus gaimardis*, similar in size and appearance to the hemigrapsid shore crabs in Bodega Harbor. Now, with the incursion of the green crab, the population of paragrapsids seemed to be falling. Ruiz presumed the green crab was responsible—"We see green crabs cruising around with *Paragrapsus* legs hanging out of their mouths"—but as in Bodega Harbor, he was set to test the connection beyond mere anecdote through a series of field experiments similar in design to the ones in California. Already he, Grosholz, and the younger biologists had erected thirty small wire-mesh crab cages on the sandy floor of the Falmouth lagoon, but now, at high tide, they were too far underwater to see.

All crab experiments require crabs; the aim of today's outing was to collect some. Ruiz stepped into the shallows of the lagoon and, the rowboat still in tow, began wading toward a white foam buoy that floated twenty yards out. The water there was above his waist. The buoy marked one of several crab traps Ruiz had baited with mackerel and set on the floor of the lagoon. With two hands he hauled at the rope below the buoy and, with some effort, heaved a large crab trap out of the water and into the boat. It was teeming with green crabs: large, small, clattering, scrabbling, dripping water. The mass of them shifted almost as a single seething organism—shapeless, encrusted, with a thousand pinhead eyes. "At least we know we have predators," Ruiz said. "Now all we need is prey."

As difficult as it would be to completely eliminate the brown tree snake from the island of Guam, the prospect of containing the green crab is even more daunting. The crab is smaller, far more numerous, and in its watery habitat occupies far more square footage than the brown tree snake could ever dream of holding. The green crab spawns early, often, and in profusion; its microscopic larvae are as numerous in the water column as the catkins of a willow on a summer breeze. In New England, in California, and elsewhere, a suggestion is sometimes made that the green crab might form the basis of a profitable industry in its own right. When life gives you green crabs, make green-crab cakes. These people have never spent an hour with mallet and tweezers separating shreds of crab from shards of shell. Although some green crabs grow to half the size of blue crabs, a favored crab for canning, the vast majority are too small to be worth the tedium. Some years ago an efficiency-minded person invented a machine that shakes blue crabs with such force that meat and shell part ways, easing the process of canning. But even this device would be unlikely to make a noticeable dent in the burgeoning populations of green crabs around the world. There is also the problem of potential success. If introduced green crabs wind up supporting a thriving industry, any effort to eradicate them would be met by the industry's opposition. By the same token, if a green-crab industry could somehow eliminate the crab from local waters, what's left as a business incentive? Ruiz said, "Why would a fisherman invest a lot of money and equipment in a market that's designed to crash?"

A more controversial approach is biological control, a program with a long and decidedly checkered history in terrestrial ecosystems. In recent years scientists have had some success controlling the spread of purple loosestrife, a fast-growing wetland weed, by introducing a weevil that feeds on it. But the failures are spectacular. In the 1930s someone had the notion to introduce the cane toad, a native of Central and South America, to counter an infestation of beetles then raging through Australia's sugarcane industry. Alas, cane toads do not eat sugarcane beetles. They do eat virtually everything else, however, and have since become one of Australia's leading pests. They grow to the size of dinner plates and at night dot the roads like cow pies; their skin secretes a substance so toxic it kills even dogs that try to eat them. A recent study found that cane toads have become a local force of natural selection, favoring the survival of native snakes with mouths too small to swallow the toads.

While I was in Tasmania, the Australian newspapers were filled with news of a more recent biocontrol experiment gone amok, involving scientists on the Australian mainland that had developed a virus to control the exploding population of English rabbits. Before all the safety tests could be completed, the virus had escaped the research facility (in a mosquito) and began infecting rabbits. The good news was, the virus worked: it killed rabbits and seemed unlikely to spread to people. However, the situation presented a dilemma: unless the virus was released, untested, on a widespread scale against all the nation's rabbits at once, the rabbits would gain immunity faster than the virus would spread, and years of scientific effort would be wasted. Already, despite strict quarantine laws, several farmers in New Zealand had secretly imported the virus and set it loose there. In a sense, biocontrol can be thought of as the forward application of invasion biology, with all the inherent complexities. Advocates emphasize its "naturalness"—no pesticides or chemicals are involved—and promise a predictability in nature that their counterparts in invasion biology have yet to actually corral.

Biocontrol has not yet been attempted in a marine setting. The vagaries of the environment—precisely how organisms reproduce and spread, and how quickly they do so—are still too daunting. Yet if ever there was a candidate target, the European green crab is it. One afternoon at the marine-pest lab in Hobart, a visiting biologist from the States presented a brief talk to describe a parasite that he had begun to investigate as a potential biocontrol agent against the green crab. His own background was impressive and included an important advance against the worm that causes river blindness in Africa. He had returned just recently from northern Australia, where he had met with scientists hunting for pathogens that might control the brown tree snake on Guam. Against the green crab, he proposed employing *Sacculina carcini*, a parasitic barnacle known to infect *Carcinus maenas* in its home range in Europe.

Although a barnacle in name, *Sacculina carcini* is unlike any barnacle imaginable. It begins its life cycle as a tiny gelatinous bleb. Eventually it finds a crab, whereupon it lands and transforms itself into what is essentially a hypodermic needle, with which it pierces the crab's body and, syringelike, injects its parasitic innards into the crab. Once inside, it sprouts roots, an extensive system of them that reaches throughout the body of the crab, shutting down the animal's reproductive system and sapping its energy. The crab remains alive all the while, moving, feeding, continuing its business, even trying to reproduce—except that the young it broods are larval barnacles, not larval crabs. The barnacle belongs to a group of similar parasites known as rhizocephalans—"root brains." *Sacculina carcini* could be an *Alien* for alien crabs, their *Night of the Living Dead*. After the presentation I came to think of the parasitic barnacle as the embodiment of the larger problem of biological invasion: the insidious internal advance, the quiet transformation from desirable to undesirable outcome. By effectively dissolving the distinction between itself and its host, the parasite obscures the crab's ability to perceive the invasion under way. It hides in plain sight. Meanwhile the host, blissfully unaware, does the work of further disseminating the invader. Imagining itself as an organism in control of its fate and the trajectory of its offspring, in fact it is gradually fashioning a future in which it is merely a vessel for another form of life.

As a candidate for the position of first-ever marine biocontrol agent, however, the parasitic barnacle still has several tests to pass. One will be to demonstrate that a relatively low initial dose of the parasite—introduced to a lagoon, say—results in a high rate of infection among green crabs. Another test, just as important, will be to show that native crabs would not succumb to the parasite. Ruiz wonders whether it will work at all. In Europe, 70 percent of green crabs are infected with rhizocephalans, yet they remain the dominant decapod in some ecosystems. "Still, it would be an interesting experiment to see the effect of a pathogen on a population. You could address a lot of interesting questions with an experiment like that."

Ruiz sat on the front steps of Miss Lyle's cottage in the late morning sun. A plastic garbage pail was within reach on the grass; at its bottom were several dozen green crabs of various sizes, and the pail echoed with the scratching of several hundred agitated legs. From time to time Ruiz gingerly retrieved a crab from the bucket, fixed a loop of fishing line to its carapace, then deposited the animal in a second pail. Several crabs now scuttled around the bottom of this bucket, trailing their fishing-line leashes like dogs set free in a dog run. It was the first day in several that Ruiz had not roused himself at five in the morning to check his e-mail from Edgewater, review a scientific paper, or set up a crab experiment in the nearby lagoon. The day was hot, the clouds sluggish. Along the coastline the sea tumbled in languorous turquoise rollers. It was like California, only better, and three times farther from the Chesapeake lab, where Ruiz's schedule follows a stiff mathematical rhythm. "You sleep less, get more work done, get strung out, sleep less, get work done, feel better about getting work done, sleep less, get strung out."

Now was perhaps not the best time for him to be away; just the day before, he'd received an e-mail message about a pending funding crisis. There were important meetings to attend, critical discussions about programmatics. "On the other hand, for green-crab research, this is an excellent time to be here. Maybe that's okay if I don't try to be involved in so many things. The scope of the program at SERC is quite broad. That makes sense. But any one of its components could make a full-time job. So I'm thinking now about scaling back. The other option is to go ahead, to create a sort of mega-lab. I can get a lot done. But I worry about the overhead, and my place in it—becoming more of an administrator and doing less research. I struggle with that. It's kinda weird to come halfway across the world to think about it."

He was also thinking about the two marine biologists who had come halfway around the world for a tour of the Falmouth field site. They were friendly—one in particular displayed a keen interest in Ruiz's research. Their departure left him visibly relieved. "The nature of the relationship is still undefined," he said. "For the most part, ecology is not like, say, medicine, where it's five people in a lab competing. In ecology it's more about ideas and laying claim to them. I worry a little about that. I made a lot of links he wasn't thinking about. Don't get me wrong, I don't think he's a sinister guy. It's just the uncertain nature of the relationship that concerns me. The green crab is a good model for a lot of different things. I want to be careful to make sure there's not a conflict."

One factor in the green crab's continued success as a global invader is its catholic appetite. It is scavenger or predator, as necessary. It eats snails, native shore crabs, introduced green crabs. "They're not food-limited," Ruiz said. "They're not predator-limited. They could prey on each other. For opportunists, cannibalism ranks high." It is a crab-eat-crab world, even for green crabs. This fact made itself terribly evident in Ruiz's bucket. A dozen or so male crabs were on the bottom, scuttling under and into one another. In the midst of it, one small green crab had been unable to resist nature's imperative to molt. He had grown too small for his shell and had squeezed out of it, leaving a pale, hard husk of himself resting on the bottom of the bucket. For any crab, the moment of molting is the most stressful and hazardous of its life, and one which, if the animal continues to grow, is destined to recur. With no exoskeleton to shield it, the animal is fully vulnerable to the world. This particular crab would have benefited from a less-public molting. No sooner had he shed his molt and exposed himself to the possibility of growth than his bucketmates turned on him. There was a brief frenzy of snipping and fending, and then, in a minute, he was gone, the entirety of his flesh reduced to shreds that dozens of tiny claws now rushed to cram into dozens of tiny mouths. Ruiz looked down on the carnage with a little smile.

"*Bon appétit*, little crabs!"

The currents of inquiry set in motion by Jim Carlton extend far beyond him now. They have grown with time into global gyres, bearing their navigators to the shores of ever more distant seas.

"Jim is the one who started it all," Bill Walton says. "He did some great history, and he's very good at speculation; he left a lot of pregnant questions. Then there's a sort of second generation—Greg, Ted Grosholz, Jon Geller, John Chapman, Andy Cohen. As you can imagine, there's a certain amount of competition going on. There's more at stake: they have tenured positions or labs, and families depending on them. But I wouldn't tell the second generation that they're the second generation. Then I guess there's this third generation."

Walton was referring to the large and growing company of budding marine biologists that happens to include himself. As a larval undergraduate, he spent a semester at the Williams-Mystic program working under Carlton. Later, as a graduate student in Ruiz's lab, he sampled the ballast tanks of transatlantic ships and, with the aid of a microscope, sorted through swarms of rowing copepods. By the time he reached Tasmania, he was pursuing a doctorate. Ruiz was one of his academic advisors; Carlton was one too, informally. Like Ruiz, Walton had begun to follow the green crab around the world. After leaving Tasmania's Falmouth, he returned to the States, went to another Falmouth—this one at the elbow crook of Cape Cod, Massachusetts—and took the ferry over to Martha's Vineyard to collect a summer's worth of data for his dissertation on the continuing impact of the green crab on the local shellfish industry.

His housing was based on a grad-student salary, which on the tony Vineyard amounted to an old cottage behind a Kiwanis meetinghouse a mile or two outside of the town of Oak Bluff. The yard was littered with the tools of his trade: crab cages, buoys, rolls of black wire mesh, oyster shells, a pair of waders hanging upside down from a peg, a wet suit on a clothesline, a snorkel, sieves. A bicycle hung on the outside wall under a plastic tarp. Inside, in the shack's single room, Walton kept for company a laptop and printer, a woodstove, two baseball mitts, a pair of cleats, several maintenance guides to motorcycles, books about food webs, a handful of postcards tacked on the wall, and, near them, a large classroom map of Tasmania. A statesmanlike gray-and-white cat named Barkeley perched on the kitchen table. Barkeley had accompanied Walton through all the stages of his educational career and was known to Walton's peers as "the cat with three degrees." Although the scholar attempted to strike up a conversation with his visitor from New York, the language barrier proved insurmountable.

On Martha's Vineyard, where the green crab has been entrenched for nearly two centuries, the local oystermen and clammers have long been engaged in an informal program to reduce the invaders' numbers and so protect their own take. Historically this has involved trapping green crabs by the bushel and carting their carcasses to the dump. To the best Walton could estimate, this effort has had little effect beyond exercising the shellfisherman. But until he came along, nobody had taken a hard, data-driven look at the effectiveness of their labor. "What is the control program doing to the green crab? People keep taking out green crabs, but why is it a forty-barrel catch each time? Shouldn't the catches drop off? You could stop trapping—but then how much worse would the problem be? Over what area can you decrease the catch and still have an impact on the population? Over what period of time? Can you reduce part of the population—say, skim off the largest ones? Finally, what happens when you do control the green crab? What happens to the quahog population? To the food web?" Walton likes working with experiments, and he likes the practical aspects of fisheries research; he likes people. "Being just another scientist at another university wouldn't be enough for me. I want to solve some problems. A lot of basic science won't ever really be used. But for me personally, I need to do something that I can see applied in my lifetime."

The time comes in the life cycle of every drifting barnacle larva when it must choose a substrate on which to settle down: the skin of a whale, the hull of a ship, the shell of a green crab, the possibilities determined by the species of the barnacle involved, an inborn fact over which the organism has no control. A young marine biologist brings more free will to the matter, which can be stressful in its own right. To Walton's dismay, he cannot easily pursue both field ecology and fisheries biology simultaneously; the professional tide pulls one direction or another. Fisheries biologists have little time for field experiments and the testing of ecological theories; there is hard, real-world work to be done. Academic marine-biology departments discount applied research as menial, blue-collar, lacking in intellectual content. They do not want to see the phrase "applied fisheries" on a job application form regardless of the quality of the work. "Those are double zeros on the cash register of academia," Carlton gently advised Walton one evening. "I've been on a lot of hiring committees. The first thing they look at is publications. If they see the word *applied*—boom." He made a chopping-tomahawk motion in the air with his hand. Ruiz has encouraged Walton to take the experimental path—to aim for the prestigious journals, to publish high-profile papers laced with theoretical models and mathematical tables, to attract the big grants.

Lately, in addition to his green-crab research, Walton had taken a more than academic interest in a native species, *Placopecten magellanicus*, the common sea scallop. The local shellfish hatchery on the Vineyard is paid by the government to spawn and rear scallops and quahogs for the benefit of commercial fishermen and the general public. When the young seeds are ready, the hatchery carefully releases them into the wild, to grow to catchable maturity. Walton was interested in the fate of the young shellfish after release and what impact the green crab might be having on them. It was at the hatchery that Walton first met his girlfriend, Beth Starr. Her job was to breed the scallops. Walton was eager to introduce her, so we drove to the hatchery one afternoon so that I could meet her and learn what goes on there.

Suppose it is spawning season for sea scallops, which begins in July and lasts into September or October. A hatchery worker—Starr—fills several cake pans with water and spreads them out on the counter. Pyrex trays, the kind meat loaf is made in, will also serve. Into the trays go those scallops that Starr deems to be ripe. She used to keep the male scallops separate from the female scallops, but now she says it doesn't matter, at least initially, and she puts them all together. Later, when things heat up, she will have to separate them again.

So the scallops sit in their trays, clapping if they're happy, gaping at one another with a thousand blue pinheads for eyes. This is the courting stage. Then comes . . . Well, suffice it to say that events proceed only when the temperature is just right (about 19 degrees Celsius), and that it is up to Starr, standing by with the thermometer in one hand and a running garden hose in the other to regulate the temperature, to get the scallops calm and relaxed and ready. Don't slam the door, no shouting, shush. "Bill usually comes in and says dirty things," Starr says.

First the males go. They spin around in circles; they rattle in their trays; they puff out clouds of milky white stuff into the water. Sometimes this happens very quickly, Starr says; sometimes it takes three or four hours. Usually they go more or less all at once. For technical reasons, the females now are quickly separated out. And after half an hour or so, they go: they spin in circles, rattle in their trays, puff out clouds of grainy orange stuff, what looks like V-8 juice, Starr says, or maybe Clamato.

Things are happening fast now, and there are lots of scallops to watch, and the males and females really shouldn't be together at this point, so Starr is suddenly shuffling: this scallop into that tray, that scallop over here, the whole scene resembling a kind of marine mixer, a hybrid of musical chairs and an undersea dating game. Starr is a malacological matchmaker, a mollusk yenta. Afterward, the spent scallops are returned to their spacious tanks of water to await the next season. Meanwhile, the products of their enthusiasm are gathered up: a bucket of eggs, all of which will be kept; and a bucket of sperm, which will be winnowed down to a beaker's worth because that's all that's needed.

"The eggs are what matter," Walton says.

Starr says, "Sperm—that's a dime a dozen."

They mix the gametes together, and the fertilized eggs—tens of millions of them—are placed in tanks, bathed in seawater, fed with algae, doted on. Mostly it is Starr who raises them; Walton hovers around and keeps the peace. The scallop larvae eat and grow strong, and eventually, after a few weeks, they are as big as specks of salt, then as big as lentils. When they are as big as dimes, they are juveniles, scallop teenagers. When they are as big as silver dollars, they are ready to leave home. Then, on a warm autumn afternoon, Walton and Starr will put on their waders and walk out onto a mudflat that Walton has decided is just right, the black muck sucking at their steps the whole way. At the appointed place, they will drop millions of silver dollars into the turning tide, like apple seeds, and watch them disappear.

Then they will drive home. Walton will make dinner, and Barkeley the everlasting cat will yowl from somewhere, and the sea anemones in their tabletop aquarium will wink at each other. They will all spend a cozy winter there, thinking anemone thoughts, and of spring, and of Tasmania maybe. The scallops have settled in, too. There are a hundred million of them or more out there at this moment: bay scallops and sea scallops and quahogs and clams—molluscan seeds, fruit of the sea—a hundred million bivalves of love, waiting for this day, this moment, eyes wide, spinning in circles, clapping.

24

It may be that when humans talk about our relationship to nature, what we mean to discuss is our relationship to time.

By and large that is what natural scientists have in mind. Their outlooks fall into roughly two camps. In one are scientists who look at a community of plants and animals and ask: What holds it together? How have these organisms managed to coexist for so long? What is necessary to maintain their current joint survival? These scientists consider the roles that different species play in an ecosystem, the dynamic interactions between them, and the ineffable factors that keep the whole assembly going. Scientists of the other camp look at an ecological community and see the handiwork of countless past contingencies: chance arrivals, historical accidents, stochasticities. Whether you ascribe to one view or the other depends on the temporal scale you examine. Scientists of the near term are more likely to be impressed by the stability of ecosystems. Those with a longer view are impressed by the ephemeral nature of ecological communities—how their memberships can see enormous turnover, gradual or episodic, on timescales of centuries to millennia. The only constant is constant change; sooner or later every species goes extinct.

Both sides essentially ask the same question: Do ecological communities that formed over a geological timespan differ in some fashion—in productivity, in potential stability—from those that were tossed together last month, last year, last century? Do recombinant communities differ from "normal" ones? Does time matter?

Neither camp would claim to be fully right. Both agree that the opposing side brings some compelling evidence. What they all wonder is, Which one is more right most of the time? "This is one of the most fundamental unsolved problems in ecology today," writes the ecologist Stephen Hubbell. "Applied ecology and conservation biology and policy critically depend on which perspective is closer to the truth." The paradox is that the study of this crucial subject is itself compromised by time. For all our ingenuity, humans have yet to devise a technique for making concerted measurements of ecological communities over time periods longer than the average human life span. "Humans are only here for a short period of time, but the systems we study last much longer," Jim Drake of the University of Tennessee said over the phone. "It's an issue of snapshots." Drake described a playful, long-running argument he has with a colleague. "I've been pushing the point that there is no stability. He says, 'Good God, man, just look out the window! It's the same today as it was yesterday!' And I say, we're not looking at it long enough. A hundred years is a short timescale. Go back a million years—what has persisted? Long-term ecological research to the National Science Foundation is five years, maybe ten or fifteen. That's just not long enough. That's the dilemma. Astronomers have gotten around that. We may never have answers to those questions, because of humans being a blip on the scene."

Therein lies the value of unadulterated nature, wherever one can find such a thing, Jim Carlton contends. Human scientists have not yet had time to determine if time matters. And until such a time comes when time is deemed indisputably irrelevant to the structure of nature, it pays to be prudent and keep some unspoiled nature around. "So I find out that this thing arrived in 1616. It's on dock pilings—so what? Why should you care? It gets to the heart of our concerns. We want to know how the natural world is assembled. The heart of ecology is to understand evolution and evolutionary biology. You can't interpret the results of evolution if you can't distinguish natural history, how our world is naturally constructed, from human-mediated history. But that's a very academic reason."

Carlton was at the helm of a rented minivan, navigating the twists and turns of the John Muir Parkway along the upper reaches of San Francisco Bay, north of San Pablo, not far from the town of Hercules. Another expedition of marine biologists was nearly complete. From the navigator's seat, Andy Cohen said, "It seems to me there are three basic reasons why people, or we as a society, might care about invasions. One is Jim's metaphor of roulette: We don't know which species is going to come in, or whether it will cause an economic problem or affect some part of the environment that we care about. Our ability to predict is nil for all practical purposes. So we have to guard against everything. Second, there's a scientific reason for caring. Ecologists try to understand how our world became the kind of world it is and how it functions as a natural system. That assumes that a big ecological and evolutionary adjustment has happened over time: that native organisms have developed relationships that governed their relative population size and distribution—what's here and what isn't, and how that changes seasonally. Now, take some organisms and throw them into the system. All of a sudden you can't see that anymore. I mean, we don't learn much about how natural systems work by going out to a golf course and doing a study.

"And that gets to the third point, which is essentially an aesthetic one and in many ways the most important. The wholeness and beauty of the natural environment is something that means a lot to people throughout the world. Not to everybody, but clearly to a large number of people. And as the world becomes man-made, I think the human spirit really suffers. In a subtle way. We lose the sense that there's a world outside of us, beyond us, that creates the sort of beauty in nature that isn't produced by us doing the landscaping or governing what's there and what isn't. It's not an either-or thing. It's a gradient: we can have it more like that or less like that. And I think it's worth making some effort to have more of it, to have more nature out there. Nature is— What was that!?"

He looked out suddenly at the highway ahead. The car in front of us had hit a pigeon and a cloud of white feathers exploded off its windshield. Our lane momentarily filled with tufts of down, which scattered and swirled away as the van sped forward and into the clear again.

"Oh, sick. One bird gone. As I said, nature is a hard path to hoe." Cohen gave a wry laugh.

After a moment Carlton said, "It seems to me that the problem of alien species has to be addressed almost as a pragmatic, target-user question. A subset of the people who won't care about the issue may be a disappointingly large number of people who find relaxation, enjoyment, peace with nature in an artificial setting. I mean, how many people go to Hawaii on vacation and enjoy it? You land there, you might even get to some slightly remote place, you lean back, you've got a drink in one hand, you're thinking of your office in Chicago, and you're saying, 'Ah, here I am: tropical vegetation, birds tweetin' around me!' And everything around you is introduced. Not a single plant, none of the lowland birds in Hawaii are native. So what is your perception of a 'world of nature'?

Not a natural world, but a world of nature that satisfies people's needs apart from the humdrum existence of their life. And it looks pretty nice. Indeed, I suppose ninety-nine percent of the tourists who visit the Hawaiian Islands don't know that the lowland vegetation and birds are introduced. 'It's tropical, for God's sake, that's why I came here. Don't bother me.' That's why I think the concept of 'biological pollution' is a tough sell. Biological pollution means, I don't know, a badly polluted river or something. The oyster industry of the Pacific Northwest is based on exotic species. Nobody in that industry thinks the oyster is really a polluting organism. Has it had an ecological impact in the Pacific Northwest? I suppose so. But it's a multimillion-dollar industry. It employs a lot of people."

I noted aloud that for many residents of Manhattan, perhaps the most cosmopolitan island in the world, nature begins and ends at Central Park. That was Frederick Law Olmsted's intent in the mid-nineteenth century when, with Calvert Vaux, he designed the park: to capture an essence of wildness and place it at the center of urban life. "I go to Central Park in the summer," I said, "and it feels like nature—if I've never been anywhere else."

"How many of the species that you're looking at are native?" Carlton asked.

I had to confess I had no idea. More than zero, certainly. I know a cedar waxwing when I see one, a blue jay when I hear one. I've seen red-tailed hawks soaring on thermals above the Metropolitan Museum of Art and peregrine falcons nesting in the ornamented balconies of tall buildings. A few years ago scientists from the American Museum of Natural History discovered a new, native species of centipede in Central Park. I've also read about the colony of escaped parrots that nests in electrical transformers in Queens and occasionally threatens the public power supply. Once, in Riverside Park, I saw a wild turkey.

I said, "Some are."

"It doesn't matter to you, though, does it? You're relaxed," Carlton said.

Cohen was baffled. "Central Park feels like nature to you?"

"It does to people who grew up in New York City," I said. Evidently John Muir found it a little too wild. He had heard of Olmsted's famous creation, the paragon of designed urban wilderness, and expressed a longing to see it. But when at last he visited New York City in 1868, he changed his mind and stayed close to his ship. "I saw the name Central Park on some of the street-cars and thought I would like to visit it," he wrote. "But fearing that I might not be able to find my way back, I dared not make the adventure. I felt completely lost in the vast throngs of people, the noise of the streets, and the immense size of the buildings. Often I thought I would like to explore the city if, like a lot of wild hills and valleys, it was clear of inhabitants."

Carlton said, "I think people do appeal very strongly to not changing their current social, economic, or aesthetic state. It's a visceral thing. The zebra mussel is a problem with a lot of people in the Great Lakes, not because it's filtering out the lakes or changing the ecosystem, but because . . ."

"It changed what they knew," Cohen said.

"It changed what they knew," Carlton said. "Most people in Europe don't know it isn't native. It is a pest species in some ways, but they don't know it isn't native. The goal is to enjoy your environment. In order to relax in the environment, you alter it to a state in which you can achieve enjoyment. Look at New Zealand: English sheep grazing on English grass under English oaks—that's all you see in some places. I think that must be incredibly satisfying."

"Hawaii, in our culture, is by and large the vision of paradise," Cohen said. "And so what you have to explain to people is that paradise is . . ."

"Lost," Carlton said.

"A façade," Cohen said. "Rotten, in some sense." He thought for a moment. "That's a big fish."

Carlton agreed. "That is a big fish."

One summer afternoon not long ago, Bill Walton and Beth Starr stopped by Carlton's office in Mystic—the office where he greets visitors, the tidy one. Carlton spoke with them individually, in private, beginning with Starr. He asked her a number of questions, including, "What is Bill's favorite color?" Green, she said right away. She left the office, and Walton entered. Carlton asked him the same series of questions.

"What's your favorite color?"

"Green," said Walton.

"And what's Beth's favorite color?"

Walton stalled. He hemmed and hawed. "Purple?"

"No," Carlton said.

"Well, it's probably something bright," Walton said. "Blue?"

"Nope, green," Carlton said, and they both laughed. "Don't worry," he added. "We can never remember these things."

Among the various titles, duties, offices, and responsibilities held by Carlton, he is also a minister of the Universal Life Church, an assignment he acquired by mail order back in the '60s in Berkeley. "Those were the days," he says. He has invoked the title three times; the marriage of Bill Walton and Beth Starr marked the third occasion. The wedding was held in early October—Columbus Day weekend—on Martha's Vineyard. The combined salaries of two shellfish biologists cannot begin to cover the cost of renting the usual wedding venues on the Vineyard, so the ceremony was held in the Whaling Church in Edgartown, for just five hundred dollars. One friend read some poems by Shel Silverstein. Another, a former colleague from Ruiz's lab in Edgewater, read a love poem by Pablo Neruda.

> *. . . the sky grows downward till it touches the roots:*
> *so the day weaves and unweaves its heavenly net,*
> *with time, salt, whispers, growth, roads,*
> *a woman, a man, and winter on earth.*

A third friend described the goings-on in a shellfish hatchery: how scallops spawn, how at that very moment there must be a hundred million of them or more—molluscan seeds, fruits of the sea—a hundred million bivalves of love out there, spinning in circles, clapping. Carlton spoke of the sea and of a primordial upwelling of love and devotion. The ceiling of the church arched far overhead. Sunlight streamed in through tall windows; the walls were a pale sea-gray. For a moment, sitting in one of the wooden pews, a guest might think he was in the grand cabin of a very large sailing ship or, perhaps, in the belly of the gentlest of whales.

Afterward there was a clambake, ministered for a nominal fee by a group of Portuguese-American women from New Bedford who cook for the local annual Great Holy Ghost Feast. They served all variety of shellfish: stuffed quahogs, scallops wrapped in bacon, clam chowder, steamed mussels and littlenecks and cherrystones, and succulent raw oysters. There were all varieties of shellfish people too: researchers, managers, fishermen, constables. There was a dinghy filled with ice and beer. The cake was chocolate with white candy scallops. Several guests remarked on the notable absence that day of Greg Ruiz: he was back at the lab in Maryland, swamped with work, steeped in stress. "The last time I saw him, his hair was out to here," one guest said; she held her hands on either side of her head and imitated the appearance of a person with all hairs standing on end. Jim Carlton, who had flown in from Brussels for the ceremony, skipped the reception and headed back to Mystic immediately after the official part.

"I'm cutting back on my travel to focus on unfinished projects," Carlton said the following afternoon. He would be resigning from all international committees. He was relinquishing his post on the American ballast delegation at the International Maritime Organization. After twenty-one years, he was ceasing formal involvement with the International Council for the Exploration of the Sea. His trip to Brussels that week marked his last as chairman of the ICES committee on introduced marine organisms, a position he had held since 1990. "And I've said no to all international travel for next year," he said. "As you know, travel is erosive to productivity."

He was sitting on a sofa in the waiting room of the Williams-Mystic administrative building. A plastic globe of Earth rested on a coffee table in front of him, and his hands hovered above its stilled ocean currents. He had just come in from the local public boat launch, where he and his students examined the undersides of small pleasure craft as they were pulled from the water. "I've had the sense for a long time that small-boat traffic acts as a vector in coastal areas. But when I looked at the literature, there was nothing quantitative. Until now, the best I could do was say, 'Trust me, I know.' There's always more species diversity on a boat than you suspect." Sure enough, there were plenty of small barnacles on the boats. "A lot of them are on the propellers—whoom whoom whoom." He made a circling motion with his hand. "A lot of little headaches."

Although they have lived on the East Coast for more than twenty years now, Jim and Debby Carlton still think of themselves as Californians. In truth, they could live anywhere. His professional status is such that any marine research program or facility in the world, including the myriad universities on the Pacific Coast, would eagerly create a niche for him. Yet Carlton likes the east and doesn't see leaving anytime soon. Anyway, he says, he travels so much that he sees the West Coast about as frequently as he would if he actually lived there. A few years ago the Carltons purchased a home in Stonington, a short drive from the Williams-Mystic office. It is set amid trees overlooking a small, bright inlet bordered by reeds and well-kept homes. A new dock extends from the lawn a dozen or so feet out over the water, affording a top-down view of the intertidal fauna. As some homeowners proudly display their azaleas and clematis, so Carlton admires the resident marine invertebrates: busy crabs; steadfast barnacles; the occasional itinerant comb jelly, ghostly and unhurried. They are friends and neighbors. Some are settlers from abroad; as Carlton sees it, those are less cause for concern than food for thought.

The one place on Earth to which Carlton is most firmly rooted, however, is a small room on the first floor of the house overlooking the cove. This is his office away from the office, his home at home. "Everything I produce, everything I write, comes out of this hole right here." More so than any of his other offices, this one is awash in paper. Every available surface—desk, table, windowsill—is stacked with journal articles, monographs, conference summaries, books, letters, treatises, entreaties. Dozens of stacks, two or three feet high, the combined future contents of perhaps twenty or thirty Carlton-authored papers yet to be written, are suspended above the floor, mid-thought. In the middle of it all, a single swivel chair. Despite the appearance of overwhelming chaos, the stacks are neatly organized by subject, such that Carlton need only rotate to the stack corresponding to whatever paper he is currently writing, extract the article or book he needs to complete a sentence, then swivel back to his computer keyboard. "I more or less know what's in each pile. Sort of. I have about a ninety percent recovery rate: ninety percent of the time I know where to find what I'm looking for."

Most mornings Carlton is in the swivel chair by five o'clock—filtering, writing. To the Seaport office by nine, home at five, back in the chair by eight in the evening. A never-ending tide of e-mail rises on him. Although Carlton himself does not read novels, he has heard that some of his colleagues do; he hears this from Debby, who talks to his colleagues on the phone. He wonders aloud, "How much of your personal recreational time do you devote to keeping up with your career?" It was a rhetorical question. "To be honest, if I'm not sleeping, I'm working. I try not to sound like a complete nerd about it. Nobody tries to be a workaholic, but from the exterior, they still look like one." Carlton does buy books, nonfiction ones related to marine science or maritime history. He estimates that he adds twelve horizontal inches' worth of books to his shelves each month, twelve new feet of books every year. From a nearly filled shelf in the hallway outside his home office, he showed off the latest acquisition: an encyclopedic volume, published by an Italian press, of the marine invertebrates of Japan. The pages were filled with small type and lavish illustrations and photographs of sea stars, sea worms, amphipods, shrimps. "It's beautiful," he said in wonder and apology. An empty box from an online bookseller sat on the dining-room table. The family cat had claimed it for a nest; when nudged, she uncurled slightly in drowsy irritation, then fell back asleep.

What does not fit in Carlton's downstairs office goes upstairs in a room roughly twice the size, also heaped with paper: the spillover reservoir. What cannot fit in this room he lugs to campus and deposits in his office in the basement of Kemble House. Some years prior, when I first visited the Kemble office, it seemed inconceivable that it could hold any more musty books or heaps of papers than it already did. Today in Kemble there are heaps and stacks so tall that even standing up, one cannot see over them. Although Carlton himself almost never works there, he readily lends the door key to visiting researchers. Soon Kemble will be tested to its limit. Debby is preparing to renovate the Stonington house: her studio and both his home offices will be torn apart and rebuilt; all his scholarly effects will need to be moved to Kemble temporarily. Carlton considers the move with the anxiety of a crustacean considering its next molt.

If the sum contents of Carlton's various offices and libraries do not yet constitute the largest collection of maritime literature in the country, if not the world, it may soon. He lends rare items to other researchers indefinitely or, if he knows he'll never need them, permanently. I wondered aloud what will become of the collection fifty or sixty years from now when . . . well, when. Carlton figures he won't worry about the books; they'll be easy enough for their inheritor to organize. His main hope is that someone saves and files his papers neatly. Although his students have grown up fluent in the language of the Internet, not all of them know how to conduct research with a card catalog. One recent addition to Carlton's collection, an 1841 manuscript titled "Report on the Invertebrate Animals of Massachusetts," may indeed be the first such biological survey of its kind in New England, but if the next generation of scholars does not know how to even look for it, it is worthless. In his students, Carlton works to inculcate an appreciation of paper. "I'm a big believer in larval imprinting."

Eventually the day will come when Carlton will retire to Pacific Grove, California, near Monterey, not to blend in with the affluence, but because his family has property there, and because the weather is warm. That time is years away—a decade at least, maybe two, he says. He has miles to swim before he sleeps. The project nearest to completion is a monograph on the introduced marine organisms of the Hawaiian Islands. Compiled with a marine biologist at the Bishop Museum in Honolulu, the catalog now stands at three hundred species. Carlton hopes soon to complete a similar monograph, in the works now for five years, on the marine introductions in the Pacific Northwest; his coauthor, Marjorie Wonham, a marine biology postdoc at the University of Alberta, has worked in the labs of both Ruiz and Carlton and, with Walton, once conducted a ballast study aboard a ship crossing the Atlantic. He handed off *Biological Invasions*, the "hugely time-consuming" quarterly journal he founded some years ago, to a new editor, Jim Drake, opening up a window of opportunity for his unfinished book about marine bioinvasions. Last but not least is the opus: a monograph on the introduced marine species of New England. Begun in 1979, it sits buried in the stack of unfinished projects that await his attention. "Our backyard is always the last thing we get around to," he mused. "We always know more about the Galápagos Islands than what's behind the marine biology building. Exotica calls."

And he still holds hopes of finding a coeditor for *Light's Manual*: a young biologist, an eager generalist. From a back corner of the Williams-Mystic office he retrieved for me a journal article he had recently coauthored with one budding candidate named Martha Hill Canning. Martha was not yet college age. She was still a nauplius, not even a cyprid yet. She was the daughter of one of the administrators in Carlton's office and was known to the staff as M-11, because that was the way she liked to refer to herself. "Her name is Martha, she was eleven years old, so she decided to call herself M-11," Carlton said. The paper was entitled "Predation on kamptozoans (Entoprocta)" and concerned her observation that certain flatworms of the genus *Plagiostomum* prey and feed on *Barentsia benedeni*, a common fouling organism in the Mystic estuary. "Predation on kamptozoans by some members of two phyla, Mollusca and Platyhelminthes, is thus now known," the paper concludes. It was published in the journal *Invertebrate Biology*. Martha was listed as the lead author.

"She's much older now," Carlton said. "She's thirteen."

A shrimp doing a headstand, antennae to the ground, legs in ceaseless motion: filtering, gathering, consuming, digesting, moving in place, broadcasting across the waves.

New World

The search for extraterrestrial life begins, and perhaps ends, in a white, gymnasium-size room in the smoggy foothills of Pasadena, California, on the sprawling campus of NASA's Jet Propulsion Laboratory. This is the Spacecraft Assembly Facility, where interplanetary probes are assembled and tested before being launched toward their various cosmic destinations. The Mars Pathfinder rover, which in 1997 captured stunning photographic vistas of the Martian surface, was built here. Spirit and Opportunity, the two rovers that in January 2004 began roaming and prodding Mars for evidence of water, were built here. Cassini, which will orbit Saturn through 2007, and Huygens, a small probe that Cassini recently dropped into the atmosphere of Saturn's moon Titan, were built here too. The Spacecraft Assembly Facility is a gateway, truly a portal to the rest of the universe. What passes through it promises to reveal a great deal about the origins, and possible fate, of life in the cosmos.

Come on in. First, however, one must be decontaminated. A visitor places one foot, then the other, into an automatic shoe scrubber, a box on the floor with spinning bristles that flagellate the soles for a minute or so. A JPL guide provides blue paper booties to slip over shoes; a blue shower cap to cover hair; and a white gown, made of paper with a shiny cling-free coating, to wear over one's clothing. Finally, an air shower: a glass booth with several nozzles blowing furiously. Then and only then, ruffled but purified, may we enter. Inside the facility, a company of blue-bootied, shower-capped, paper-gowned technicians fusses over the skeletons of spacecraft-to-be. The room is as arid as a desert, the humidity a drastically low 42 percent. The floors are regularly scrubbed to remove dander and bacteria. NASA's intent is to create an environment that is hostile to any microbes that might hitch a ride aboard the outbound spacecraft yet is benign to the human engineers who must assemble these delicate vehicles. If that sounds like an impossibility, it is. Welcome to the paradox of planetary protection.

In 1967, inspired by a new international outer-space treaty, the space-racing nations of the world agreed to spare no effort in preventing the spread of organisms from one moon or planet to another. At NASA, this mandate evolved into an official planetary protection policy, a Sisyphean effort to shield the universe from the people exploring it. Traditionally, the assumed beneficiary of planetary protection has been the planet Earth. We have all seen the movies; we know the disaster scenarios: extraterrestrial spores return from outer space, and in no time the citizens of Earth are heaps of dust or brain-dead zombies. Accordingly, NASA has developed an elaborate quarantine protocol to handle soil samples retrieved from other planets. Comforting perhaps, but statistically of marginal value. Contagion spreads from the haves to the have-nots, and so far as scientists have yet determined, Earth is the only planet with life to give. Besides, virtually all the spacecraft that leave Earth depart on one-way missions: they drift eternally through interstellar space, or they burn up in foreign atmospheres, or they sit on planetary surfaces, never rusting, transmitting data until their batteries fade away. Among all the lawns in the cosmos, ours is the one with dandelions, and the wind is blowing outward.

No, if anybody should be worried about biocontamination, it's our planetary neighbors. In the coming decade, NASA has scheduled no less than four major missions to Mars to grope for hints of water or life. Down the road is a robot that will drill below the icy surface of Jupiter's moon Europa to probe a briny ocean believed to exist there, and the Titan Biological Explorer will plumb the atmosphere of Titan for the chemical precursors of life. Interplanetary traffic is picking up, and NASA would like to avoid going down in history as the agency that accidentally turned the Red Planet green with life.

But the true worry isn't ecological; it is epistemological. Any earthly contamination—of the Martian soil or, more immediately, of the instruments sent to study the soil—would seriously muddy the multibillion-dollar hunt for extraterrestrial life. As Kenneth Nealson, a University of Southern California geobiologist and JPL visiting scientist, recently told the journal *Nature*: "The field is haunted by thinking you've detected life on Mars and finding that it's *Escherichia coli* from Pasadena." As it turns out, that fear is well founded. Not only does microbial life survive in the Spacecraft Assembly Facility; in some cases it thrives there. There is no question whether we're exporting life into the cosmos—we absolutely are. What's left to determine is exactly what kind of life is emigrating and how far it is spreading.

"Bugs are very clever," Kasthuri Venkateswaran says with affection. "They started out on Earth 3.8 billion years ago, when nothing else was here!"

Venkateswaran—bow tie, oxford shirt, smart round glasses—occupies a bunkerlike office a couple hundred yards up the hill from the Spacecraft Assembly Facility. Unofficially, he is an astrobiologist, a job description recently coined at NASA to describe the cadre of scientists involved in the agency's accelerating search for life beyond Earth. Officially, he is the senior staff scientist of the biotechnology and planetary protection group. While his celebrated colleagues design ever more inventive spaceships and robots to scour the surface of Mars for some signature of life, Venkateswaran quietly examines the machinery itself, searching for any clever microbes—"bugs," he calls them—that might try to tag along. Neat and kindly as a country doctor, he is in fact the biological protector of the universe. To colleagues and, at his insistence, visitors, he is simply Venkat.

"The life-detection techniques we have today are incredibly sensitive," Venkat says. "A few molecules could jeopardize the sample you're bringing back." He pulls out an official pamphlet: *Biological Contamination of Mars, Issues and Recommendations.* The surfaces of outbound NASA spacecraft and instruments, it declares, should be rid of living stuff, dead stuff, parts of dead stuff, and any stuff that might be mistaken for any of the aforementioned stuff. The effort is under constant review and revision. Recently NASA stopped using cotton swabs in the cleaning process: to a life-detection instrument, the atomic bonds in a stray filament of cotton could be mistaken for the signature of proteins. The last thing Mars scientists want to discover is that Martians are the evolutionary descendants of Q-tips.

In the old days, ridding the average spacecraft of bugs was a simple matter: place it in an oven, heat it up to a jillion degrees or so, and bake it for a couple of days. Today spacecraft are far more sophisticated and fragile, made of lightweight polysyllabic polymers and stuffed with microcircuits and light-years-beyond-Microsoft software. "Nowadays, most electronics can't take that kind of heat," Venkat says. Instead, the individual components of the spacecraft are swabbed down with alcohol during construction; the components that can take it also undergo some sort of heat treatment. (The swab approach is by no means bug-proof. Venkat has found that the alcohol sometimes breaks apart microbes and glues their innards to the spacecraft; this kills the microbe but leaves the prospect of life-detection even muddier than before.) The various parts of a given spacecraft are built, and decontaminated, by subcontractors around the globe. NASA readily concedes that it is physically—or at least financially—impossible to remove every speck. Instead, the agency issues guidelines intended to minimize the risk of contamination: no more than three hundred specks per square meter, say, for a landing pod actively involved in the life-detection process. The components are then sent to the JPL or another NASA campus for inspection and final assembly. This is where Venkat's research begins in earnest.

I toured the facility in the company of Victor Mora and Jesse Gomez, two of the space-age custodians responsible for keeping the place tidy. Spacecraft parts that come into the room are relatively free of microbes to begin with, they said. All that's required is to keep the density of free-floating particles to a minimum. Dust, hair, the sloughed-off skin cells of NASA workers—all are contaminants in their own right and, more important, nutritious meals for whatever microbes might be around. "We're shedding all the time," Mora said. "Even our eyes shed." Giant fans in the ceiling, several dozen feet overhead, suck particulates upward and outward into exile. The antistatic robes worn by technicians funnel personal particles down toward the floor, which is swabbed regularly.

"Microbes need particles to attach to," Venkat says. "Without particles, without nutrients, the environment is essentially extreme."

If astrobiologists have learned anything, however, it's that almost no environment is too extreme for life. In the past few years, scads of extremophile organisms have been discovered thriving under conditions once considered inhospitable. Clams have turned up in the sunless, high-pressure depths surrounding seafloor vents. Algae in the Antarctic, where conditions resemble the dry valleys of Mars, spend much of their lives desiccated and drifting in the wind, waiting for their situation to improve. Microbes have been found miles underground in hot geysers, in gold mines, in solid volcanic rock, deriving their nourishment from sulfur, manganese, iron, petroleum. In recent years, a whole new field called geomicrobiology has sprung up precisely to study tiny creatures that are otherwise indistinguishable from rocks. Astrobiologists agree that if there is life to be found beyond Earth, it almost certainly will be very small and equally hard to discern.

Trained as a microbiologist, Venkat brings to his task an impressive history of sleuthing out wily tiny critters. In 1998 he discovered a bacterium that survives the high salinity of California's Mono Lake by living *inside* the lake's rocks. After prominent newscasters and government officials were mailed anthrax spores in the autumn of 2001, Venkat published a paper later used by the Department of Homeland Security on how to distinguish anthrax from other microbes. None of his encounters in the microworld, however, quite prepared him for the discoveries he has made in Pasadena. Using a sophisticated array of life-detection methods—the same methods being refined for the hunt for extraterrestrial life—Venkat has discovered a plethora of bizarre microbes thriving in the Spacecraft Assembly Facility, microbes that would have escaped detection by older technologies. He held up a red-capped vial for me to see. Inside, invisible in a thimble-size sea of clear liquid, were the newly found inhabitants of Planet NASA. His is a true microcosm: a new world, hitherto unexplored, as enlightening as any that Venkat's stargazing colleagues will ever hope to find.

Thus far, Venkat has identified twenty-two species of microbe in the Spacecraft Assembly Facility; in other, similar NASA environments; and even on actual spacecraft. Many are microorganisms common to arid environments, such as *Bacillus mojavensis*, a bacterium that probably drifted in from the Mojave Desert. A handful are entirely new species. One, which Venkat has named *Bacillus nealsonii* (in honor of Kenneth Nealson, his former supervisor at JPL), possesses two protective coats, making it a tough spore capable of surviving in the ultradry environment of the assembly facility. As Venkat discovered, the second spore coating also offers a secondary benefit: it makes the organism unusually resistant to gamma rays, a form of cosmic radiation that, in large doses, is fatal to men and microbes alike. (Earth's atmosphere screens out most gamma radiation; Mars, in contrast, is a gamma-ray frying pan.) Tough as it is, the bacterium probably is not unique to NASA. The world of undiscovered microbes is vast, and Venkat suspects that *Bacillus nealsonii* also resides outside the assembly facility.

Venkat has found bugs in the spacecraft-assembly facility at the Kennedy Space Center in Florida; on hardware and in drinking water from the International Space Station; in circuit boards destined for an upcoming mission to Europa; and on the metal surface of the Mars Odyssey spacecraft, which has been orbiting Mars since October 2001. While Odyssey was being assembled at the Kennedy Space Center, Venkat isolated a new species of bacterium—*Bacillus odysseyi*, officially—that carries an extra spore layer, or exosporium, that makes it several times more resistant to radiation than other spore-forming microbes found in the facility. "It carries novel proteins as a sunscreen," Venkat says. Like *Bacillus nealsonii*, *Bacillus odysseyi* may turn out to live elsewhere besides its assembly facility. But what's notable, Venkat says, is that the very traits that render these bugs impervious to decontamination also grant them a decent chance of surviving the radiation shower they would encounter en route to and on the surface of a place like Mars.

One discovery, a bacterium named *Bacillus pumilis*, has given Venkat particular cause to marvel. He found the microbe thriving directly on spacecraft surfaces, presumably drawing its energy from ions of trace metals like aluminum and titanium. "Aluminum is toxic," Venkat exclaims, baffled. "There are no nutrients. There is no water." In addition, the species exhibits a remarkable defense against desiccation. The individual cells form protective spores, which then band together to create what Venkat calls an igloo. In microphotographs, this spore house looks rather like a macaroon. Moreover, when Venkat cuts open the igloo, he finds no visible trace of the individual spores; they've all dissolved into the collective structure. High-tech methods of life detection reveal no evidence of life. Yet when Venkat warms up the igloo and adds a little moisture, *Bacillus pumilis* again springs into being. If the microbe is any indication of the sort of life that awaits discovery on Mars or elsewhere, he says, good luck to the robot sent to detect it.

Bacillus pumilis itself isn't a new species. It has been studied throughout the world for years, but its igloo-forming habits were not well known. For instance, its attachment to aluminum is novel. Last month Venkat published a paper claiming that the SAF version of *Bacillus pumilis* is in fact a new species after all—a sub-strain that has adapted and evolved to the conditions imposed on it by NASA, like an herbicide-resistant dandelion or the supertough microbes that sometimes spring up in hospitals. He has named it *Bacillus safensis*, and it represents precisely the kind of organism that his fellow astrobiologists are looking for in outer space. It's not a Martian, but in form and function it may turn out to closely resemble one.

It is, in any event, one step closer than any other earthly creature to becoming the first organism to survive on another planet. Venkat has found the bacterium in every other NASA assembly facility he has studied. Three years ago he found it on the Mars rovers Spirit and Opportunity, then under assembly at JPL. At this very moment, the rovers are actively poking around in the Martian dirt, as they have since early 2004. *Bacillus safensis* is almost certainly aboard them, alive and well, Venkat says. "They could be there for millions of years, because they are spores. Whether they will become active and begin terraforming—that research is still ongoing."

The Space Assembly Facility is a standing paradox. Through its assiduous effort to avoid spreading life throughout the cosmos, NASA has created an environment that inadvertently fosters the very kind of life it is traveling so far beyond Earth to find. As Venkat says, "We have a kind of survival of fitness." What began as a means to an end is now an end in itself; the doorstep has become a laboratory, a nursery even, a small-town study in life's cosmic persistence. It is a study, too, in the impossibly high cost of perfect hygiene. In attempting to identify where some of the microbes were coming from, Venkat found that in at least one instance, they appeared to have been introduced during the cleaning process—news to make one nostalgic for the good old days when it sufficed to turn on the oven and bake everything. Wherever humans go, it seems, we go with company. Looking around the assembly facility with Mora and Gomez, I saw a man-made cosmos, every aluminum surface a habitable planet, its ethers traversed by micronauts riding spacecraft named *Human Hair* and *Eyeball Cell.*

"People are the dirtiest things around," Gomez said.

"Yeah," said Mora. "We're the contaminants."

As NASA's search for extraterrestrial life advances, it more and more resembles a trip through a hall of mirrors. The farther from Earth our gaze wanders, the more our very presence seems to nag us. Can we search for foreign life without contaminating it with our own? Can we discern the contamination from the real thing? If ultimately we are related, if we are all evolutionary relatives from way back, is there even a difference? Some scientists wonder whether logic even permits us to find anything but ourselves out there: our understanding of what constitutes life is shaded by what we know on Earth, so that's all that we know how to look for. It's like that old joke about the guy who hunts for his keys under the lamppost because that's where the light is. Venkat, in his office, nods vigorously in affirmation.

"Maybe it's something you're not able to detect with the naked eye. Maybe it exists on a different wavelength," he says. A public-relations minder from NASA had joined us; she looked less than thrilled by Venkat's speculations. He went on: "You might think I'm crazy. Maybe there's somebody walking around right now whom we can't see."

The hunt for extraterrestrial life marks the ultimate test of humankind's self-knowledge. We cannot find and recognize "other" until we can first, at the most basic cellular level, recognize "us." Therein lies the true value of Venkat's microbes. Having found *Bacillus nealsonii, Bacillus safensis,* and their kin—having in some sense fostered their creation and survival—NASA has no plans to destroy them all. On the contrary, Venkat intends to keep them alive as a sort of microbial archive for future reference. Someday, maybe soon, scientists will flip over a rock on Mars or Europa or somewhere out there and claim the profound, the first-ever discovery of "them." How to tell for certain? We will hold up a mirror and compare appearances; that mirror awaits in Venkat's office. His microbes are us: our emissaries, our representatives, the reflection of our wily selves. Deciphering and confirming the distinction—them or us—will probably take years. But as Venkat sees it, those are precisely the hard facts that humans evolved to tease apart.

"It's tough," he says. "But that's where our intelligence comes in."

Acknowledgments

This book could not have been written without the generous participation of its key characters, and of the many more who go unmentioned; I am deeply grateful to them all. My thanks also to the Bishop Museum, the Centre for Research on Introduced Marine Pests, SeaRiver Maritime, Inc., and the Nature Conservancy of Hawaii. This project received essential support from the Alfred P. Sloan Foundation, the New York Foundation for the Arts, the Virginia Center for the Creative Arts, the Writers' Room, the Yaddo Corporation, and especially the MacDowell Colony.

I could not have reached the finish without the help of many people near and dear. My warm thanks to Andrew Albanese, James Drake, Will Gillham, Rachel Kadish, Daniel and Jean Mackay, and Chuck Siebert, for offering critical feedback; Chris Jozefowicz, for expert fact-checking; Ilena Silverman and Larry Brown, for their encouragement and perspective; Jack Rosenthal, Adam Moss, Gerry Marzorati and *The New York Times Magazine*, for publishing an early incarnation; Stephen Petranek and the staff at *Discover*, for their ongoing support; Marilyn Beach, Joseph DeMeyer, Chris and Sarah Sherwood, Kim and Donald Jurney, and Ethan Watters and the folks at the Grotto, for providing sanctuary; Flip Brophy and John Glusman, for their faith and guidance; and the families Burdick, Dominus, King, Sanchez, and Broude. I am blessedly indebted to Darcy Frey, who went above and beyond. Most of all, thank you to Susan, for the happy ending.

Index

Note: A bibliography, a study guide, additional resources, and suggestions for further reading are available on the author's Web site: www.aburdick.com.

Page numbers in *italics* refer to illustrations.